PETERSON FIRST GUIDE TO
ASTRONOMY

SECOND EDITION

JAY M. PASACHOFF

STAR MAPS BY
WIL TIRION

CONSTELLATION PAINTINGS BY
ROBIN BRICKMAN

HOUGHTON MIFFLIN HARCOURT
BOSTON NEW YORK

Copyright © 2014 by Jay M. Pasachoff

Constellation paintings © 1988 by Robin Brickman

www.hmhco.com

Library of Congress Cataloging-in-Publication data is available.

ISBN 978-0-544-16562-5

Printed in China

SCP 10 9 8 7 6 5 4

4500681767

So much is going on in contemporary astronomy, with thousands of candidate exoplanets discovered around other stars, with the universe's expansion accelerating instead of slowing down, with spacecraft monitoring various levels of the sun every few seconds, and so on. But the best thing about astronomy remains going outside on a clear night, finding a dark place far from lights, and looking up. The beauty of the stars, the planets, and the Milky Way are for all to enjoy. It is fun for me to be able to describe observational astronomy in this little book and to provide Wil Tirion's maps and Robin Brickman's constellation drawings, as well as a variety of photographs from amateur and professional astronomers and spacecraft, to help everyone enjoy the sky above.

Jay M. Pasachoff

The sun and moon, the stars and planets, and the other objects in the sky are always with us. Wherever we live, wherever we travel, we can always call upon the companionship of these familiar things. No equipment is necessary for you to see them, but a knowledge of what you are looking at will add to your appreciation and enjoyment.

Have you ever tried counting the stars? In a dark place on a clear night, with your unaided eyes, you can see about 3,000 stars. From a city street, with streetlights polluting the sky with their brightness, you may see only a handful. As the Earth turns, the stars appear to move. The speed at which they appear to move depends on where they are in the sky. The fastest of them appear to move, every hour, across an arc of sky covered by the width of your fist at the end of your outstretched arm. Over a few hours, the appearance of the sky changes drastically. And from season to season, as autumn changes into winter or spring into summer, different stars become visible in the evening.

To show what is in the sky as it changes through the night or through the year, this book contains 24 star maps (see pp. 20–43). Each month of the year has its own pair, one map to be used when you face north and the other map when you face south. The maps are meant for observers at midnorthern latitudes.

The maps show stick figures for the constellations, the arrangements of stars described long ago and often associated with myths and legends. The pages following the maps show some of the most popular and easy-to-see constellations in more detail. For other topics, see:

This book is meant for naked-eye observing. If you have binoculars, some fainter objects will become visible, a few of which are listed here. If you have a telescope, you can look at several of the objects in this book in more detail. See pp. 119–121 for tips about using binoculars or telescopes.

How to start using this book? I would suggest that you begin by reading the first 19 pages. Then, on the next clear night, find the star maps (pp. 20–43) that match the date and time. Match the constellations on the map with the stars in the sky. Read about a few of those constellations in more detail (pp. 44–67), and carry on from there.

Enjoy the sky! Enjoy astronomy!

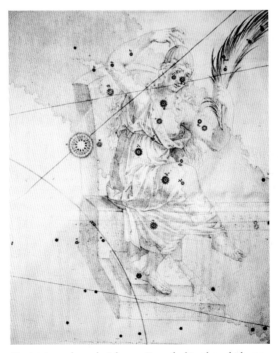

Cassiopeia, as drawn by Johannes Bayer for his atlas, which was published in 1603.

Each night, many stars appear to rise above the horizon, travel across the sky, and eventually sink below the horizon. Others remain visible all night long. If you face east, you will see stars rise in the night sky. They won't be moving fast enough for you to see their motion, but if you look back at the sky after a while, you will see they are in different places.

Partway up the sky in the north is a point around which the stars appear to rotate. This imaginary point, the *celestial north pole,* appears never to move. It lies directly above the Earth's North Pole. Halfway between the celestial north and south poles, lying over the Earth's equator, is the *celestial equator.*

If we were at the Earth's North Pole, the celestial north pole would be right above our heads—at the point called the *zenith* (zee'nith). Wherever you are, the zenith is the point in the sky directly above you. As we move from the North Pole toward the equator, the celestial north pole appears lower in the sky until, when we are at the equator, it is on the horizon. The height of the celestial north pole above the horizon corresponds to the lati-

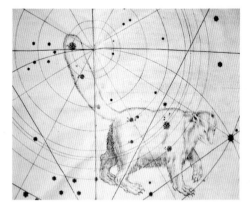

Ursa Minor, the Little Bear, as drawn by Bayer, with the North Star, Polaris, labeled with a Greek alpha and given a large symbol, even though there are 48 stars in the sky brighter than it. It is within a degree of the point around which the stars appear to rotate, the celestial north pole.

tude where you are observing. At the 40° north latitude of New York City, for example, the celestial north pole is 40° above the horizon.

All stars appear to move in circles around the celestial poles. Stars close to each pole move in small circles and never go below the horizon. We call them *circumpolar.*

Stars farther from the pole move in circles so large that part of the circles lie below our horizon. We see those stars rising in the eastern part of the sky and, later, setting in the western part of the sky. The circles are so big that the paths of the stars look almost straight.

The celestial south pole is an imaginary point above the Earth's South Pole. Neither the celestial south pole nor the stars near it are ever visible from our latitudes. Only if we go south of the Earth's equator can we see them.

Each night, the stars rise about four minutes earlier than they did the night before. Thus a given star rises an hour earlier every 15 days. As a result, we see different stars in different seasons (see p. 69). This effect occurs because the Earth rotates on its axis as it revolves around the sun. As the Earth travels $^1/_{365}$ of the way around the sun each day, it also makes one full turn (rotation) on its axis. Thus, we are facing a slightly different part of the sky at the same time each night; the stars appear to have traveled 3 minutes 56 seconds per day farther around the sky, even though it is the Earth that is actually moving in relation to the stars. We face the same stars 3 minutes 56 seconds (that is, 24 hours divided by 365 days) earlier each day.

With a multi-hour exposure, we see a panoramic view of star trails around the celestial north pole (left) and the celestial south pole (below the horizon at right).

Brightness and Magnitude

Thousands of years ago, the Greek astronomer Hipparchus said, reasonably, that the brightest stars were "of the first magnitude," the next-brightest stars were "of the second magnitude," and so on. The faintest stars visible to the naked eye were "of the sixth magnitude." Today, we still use a version of this method to describe brightness. How bright a star appears is called its *apparent magnitude*.

Modern astronomers have made the magnitude scale more precise. A difference of five magnitudes now means exactly 100 times in brightness. A star of magnitude 1.5 is between magnitude 1 and magnitude 2. Also, some objects are even brighter than 1st-magnitude stars. These stars can have a magnitude of 0, or even negative numbers. The brightest star in the sky, Sirius, has a magnitude of −1.5.

Every difference of 1 magnitude corresponds to a factor of about 2.5 in brightness. For example, a 2nd-magnitude star is about 2.5 times fainter than a 1st-magnitude star. A 6th-magnitude star is exactly 100 times fainter than a 1st-magnitude star.

Magnitudes of Selected Objects

Venus, at its brightest	−4
Jupiter, at its brightest	−3
Sirius, the brightest star	−1.5
Arcturus, Vega (bright stars)	0
Pointers at the end of the Big Dipper	2
Faintest stars on our star maps	4.5
Faintest stars visible to the naked eye	6
Faintest stars visible with binoculars	9
Faintest stars visible with a small telescope	12
Faintest stars visible with biggest ground-based telescopes	28
Faintest stars with the Hubble Space Telescope	30

How faint a star you can see depends on how much light passes into your eye. At night, the pupil in your eye can open as wide as 8 millimeters. Your brain processes a new image every $1/30$ second or so. Therefore, you can detect only as much light as comes into an 8-millimeter circle in $1/30$ of a second.

If you use binoculars, you collect light from a larger circle—the area of each front, or objective, lens. Each front lens of 7 × 50 binoculars, a type commonly used for astronomy, is 50 millimeters in diameter. Since this lens collects more light than your naked eye, you can see fainter objects when you use binoculars.

Telescopes generally have larger lenses (or mirrors) than binoculars have. Thus they collect even more light and allow you to see still fainter objects (see p. 120). Another way to collect more light is to collect it for a longer time by using a camera. Whereas the eye collects light for only $1/30$ of a second per image, cameras can make exposures for many seconds or minutes. This, too, allows you to see fainter objects.

The Brightest Stars in Our Sky
(visible from midnorthern latitudes)

Star	Constellation	Magnitude	Visible in Evening
Sirius	Canis Major (Maps 1S–4S, 12S)	−1.5	winter
Arcturus	Boötes (Maps 2N, 3, 4S, 6S, 7, 8N)	0.0	spring
Vega	Lyra (Maps 4N–6N, 9N–11N)	0.0	summer
Capella	Auriga (Maps 2N–5N, 8N–11N)	+0.1	winter
Rigel	Orion (Maps 1S–3S, 11S–12S)	+0.1	winter
Procyon	Canis Minor (Maps 1S–4S, 11, 12S)	+0.4	spring
Betelgeuse	Orion (Maps 1S–4S, 10N, 11S–12S)	+0.5	winter
Altair	Aquila (Maps 5N, 6S–10S, 11)	+0.8	summer
Aldebaran	Taurus (Maps 1S–3S, 10, 11S–12S)	+0.9	winter
Antares	Scorpius (Maps 5S–8S)	+1.0	summer
Spica	Virgo (Maps 3S–7S)	+1.0	summer
Pollux	Gemini (Maps 1S, 12S)	+1.1	spring

The maps showing bright stars and constellations appear on pp. 20–43. N = facing north; S = facing south. On some dates and times, the stars are higher in the sky than the regions shown on the maps.

ESTIMATING DISTANCES ACROSS THE SKY

To find your way from one star or constellation to the next, it is often convenient to be able to estimate angles in the sky. The distance (measured in degrees of arc) from the horizon in any direction to the zenith, the point overhead, is 90°. Your fist, held at the end of your outstretched arm, takes up about 10°. Your thumb, held at the end of your outstretched arm, takes up about 2°. The moon covers about ½°.

From our midnorthern latitudes, the Big Dipper can be seen at all times of the year. It is an *asterism*, a group of stars making up a special shape, rather than a *constellation*, one of the 88 regions into which the sky is divided (see p. 12). The stars at the end of the bowl are called the Pointers, because a line drawn through them curves across the sky almost directly to Polaris, the North Star. The Pointers are separated by about 5°, and the angle, or apparent distance, from the top star in the Pointers to the North Star is about 30°. You can judge how big an angle 30° is either by taking one-third the distance from the horizon to the zenith or by taking three fists' width across the sky.

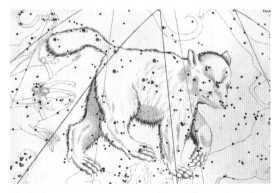

The Big Dipper is an asterism, part of a constellation, shown here as part of Ursa Major, the Big Bear. This drawing is from the unpublished star atlas by John Bevis (1749), updated and improved from Bayer's atlas.

The constellation Orion is perhaps the easiest to find in the sky when it is up—that is, when it is above the horizon, as it is on winter evenings. Three second-magnitude stars make up its belt, which is 3° long.

Two bright stars appear symmetrically above and below Orion's belt, each about 9° (almost one fist's width) away from it. Betelgeuse, in Orion's shoulder, is slightly reddish even to the naked eye. We thus know it is a cool star. Rigel, in Orion's heel, glows noticeably blue-white, indicating that it is a hot star.

In the midst of Orion's sword, which extends down Orion's leg from one side of his belt, a fuzzy region glows. This is an area of gas and dust called the Orion Nebula (p. 46), in which stars are forming. It is not visible to the naked eye, but appears faintly in binoculars or small telescopes.

Orion lies sideways, with its belt pointing down to bright Sirius and up past reddish Aldebaran to the Pleiades (p. 56). The Orion Nebula glows reddish near the belt. Castor and Pollux appear one above the other toward the image's left. See Maps 1, 2, and 3 South, suitable for winter evenings.

THE CONSTELLATIONS

People have long pretended that the sky is divided into groups of stars, each group with its own story. Each civilization has had its own legends. Many of the constellation names we use today come from the ancient Greeks. The constellations in the northern sky are in the part of the sky that was visible from the Greek empire.

When scientific expeditions went to the Southern Hemisphere a few hundred years ago, they charted the regions of the sky that could not be studied from midnorthern latitudes. The constellation names these explorers assigned to the southern constellations in the 17th and 18th centuries reflect a more modern view of the world than the Greeks had, as well as a fascination with mechanical devices.

Though many constellation names have been in use for centuries, in 1930 the International Astronomical Union agreed to divide the entire sky into exactly 88 constellations. Each star now belongs to one and only one constellation. Some constellations are too far south for us to see from midnorthern latitudes or from still farther north; they are printed in *italics* in the table on pp. 13–14.

Starting on p. 44, we discuss a few constellations in detail. Many people know the twelve constellations of the *zodiac,* the ones that form a backdrop for most of the sun's path across the sky. They are shown in **boldface**.

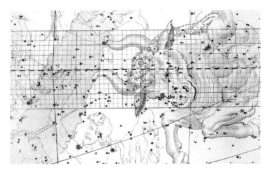

Taurus, the Bull, engraved for John Bevis (1749). Aldebaran, part of the Hyades cluster of stars, is the brightest star on its face, and the Pleiades, another star cluster, ride on its back.

The Constellations

Andromeda*	an-drom'e-da
Antlia, the Pump	ant'lee-a
Apus, the Bird of Paradise	ah'pus
Aquarius, the Water Bearer	a-quayr'ee-us
Aquila, the Eagle	ak'will-a
Ara, the Altar	ah'ra
Aries, the Ram	air'ees
Auriga, the Charioteer	oar-eye'ja
Boötes, the Herdsman	bo-o'tees
Caelum, the Chisel	see'lum
Camelopardalis, the Giraffe	ka'mel-o-par-dal'is
Cancer, the Crab	kan'ser
Canes Venatici, the Hunting Dogs	kay'nes ve-na'ti-chi
Canis Major, the Big Dog	kay'nis ma'jer
Canis Minor, the Little Dog	kay'nis mye'ner
Capricornus, the Goat	kap'ri-korn-us
Carina, the Ship's Keel	kar-eye'na
Cassiopeia*	kass-ee-oh'pee-a
Centaurus, the Centaur	sen-taw'rus
Cepheus*	see'fee-us
Cetus, the Whale	see'tus
Chamaeleon, the Chameleon	ka-meel'yon
Circinus, the Compass	sir'sin-us
Columba, the Dove	kah-lum'ba
Coma Berenices, Berenice's Hair	koh'-ma beh-reh-nee'ses
Corona Australis, the Southern Crown	koh-roh'na aus'tral'is
Corona Borealis, the Northern Crown	koh-roh'na baw-ree-al'is
Corvus, the Crow	koar'vus
Crater, the Cup	kray'ter
Crux, the Southern Cross	krucks
Cygnus, the Swan	sig'nus
Delphinus, the Dolphin	del-fee'nus
Dorado, the Swordfish	daw-rah'doh
Draco, the Dragon	dray'ko
Equuleus, the Little Horse	eh-kwoo'lee-us
Eridanus *	ey-rid'an-us
Fornax, the Furnace	four'nax
Gemini, the Twins	gem'en-eye
Grus, the Crane	grus
Hercules*	her'kyool-ees
Horologium, the Clock	haw-roh-loj'ee-um
Hydra, the Water Snake	hye'dra
Hydrus, the Water Snake	hye'drus
Indus, the Indian	in'dus
Lacerta, the Lizard	la-sir'ta

Leo, the Lion	lee´oh
Leo Minor, the Little Lion	lee´oh mye´ner
Lepus, the Rabbit	lee´pus
Libra, the Scales	lee´bra
Lupus, the Wolf	loo´pus
Lynx, the Lynx	links
Lyra, the Harp	lye´ra
Mensa, the Table	men´sa
Microscopium, the Microscope	mye-kroh-scohp´ee-um
Monoceros, the Unicorn	mon´oh-seyr´us
Musca, the Fly	mus´ka
Norma, the Surveyor's Level	norm´a
Octans, the Octant	ok´tans
Ophiuchus*	oh-fee-you´kus
Orion*	oh-rye´un
Pavo, the Peacock	pa´voh
Pegasus*	peg´a-sus
Perseus*	per´see-us
Phoenix, the Phoenix	fee´nicks
Pictor, the Easel	pik´tor
Pisces, the Fish	pye´sees
Piscis Austrinus, the Southern Fish	pye´see aws-trye´nus
Puppis, the Ship's Stern	pup´pis
Pyxis, the Ship's Compass	pick´sis
Reticulum, the Net	reh´tick´you-lum
Sagitta, the Arrow	sa´jet´a
Sagittarius, the Archer	sa-jet-air´ee-us
Scorpius, the Scorpion	skawr´pee-us
Sculptor, the Sculptor	skulp´ter
Scutum, the Shield	skyoo´tum
Serpens, the Sextant	sir´pens
Taurus, the Bull	taw´rus
Telescopium, the Telescope	tel-es-koh´pee-um
Triangulum, the Triangle	trye-ang´you-lum
Triangulum Australe, the Southern Triangle	trye-ang´you-lum aws-tray´-lee
Tucana, the Toucan	too-kay´na
Ursa Major, the Big Bear	er´sa may´jer
Ursa Minor, the Little Bear	er´sa mye´ner
Vela, the Ship's Sails	vee´la
Virgo, the Virgin	vir´go
Volans, the Flying Fish	voh´lans
Vulpecula, the Little Fox	vul-pek´you-la

Alternative pronunciations sometimes exist.
* Proper name. **Boldface:** zodiac. *Italics:* too far south to see from 40° North latitude.

THE MILKY WAY

On a clear night in a dark place, when the moon is not full, you may see a hazy white band stretching across the sky. This band is the *Milky Way*. Even with only your naked eye, you can see that the Milky Way is not regular; it has patches of white and darker regions.

The fact that the Milky Way is a narrow band in the sky tells us that we live in a flat galaxy (p. 78). Imagine that the Milky Way is shaped like a dinner plate. If we were tiny people living about two-thirds of the way from the middle of that plate, the plate would look to us as the Milky Way does. If we looked toward the edge of the plate or toward the center, we would see lots of plate; in the sky we see lots of stars and clouds of gas and dust, which are called *nebulae* (see p. 70). But looking up from the plate, we see none of it. Looking up or down from our spot in our flat galaxy, we see only the stars nearest us.

The Milky Way is shown on the star maps on pp. 20–43. On summer evenings, part of it passes nearly directly overhead, including the constellation Cygnus, the Swan, with the constellation Sagittarius, the Archer, in the south. When we look at Sagittarius, we are looking toward the center of our galaxy, which looks richer in stars, gas, and dust. Scanning the Milky Way with binoculars reveals many such fuzzy regions.

The Milky Way, seen in an all-sky display centered on the center of our galaxy, which is in the constellation Sagittarius. The Large and Small Magellanic Clouds, satellite galaxies, also show (south of the Milky Way, right of center in this display).

TELLING STARS FROM PLANETS

Though the stars appear to move across the sky in the course of the night, their positions in relation to one another remain fixed. But there are a few points of light in the sky that change position slightly with respect to the stars from night to night. These are the planets, from the Greek word for "wanderer." Telescopes reveal that the planets have different apparent shapes (Mercury and Venus have phases, for example, and Saturn has a ring) and sizes. Now that we have sent spacecraft up close to each of the planets, we have learned a great deal more about their basic natures (see pp. 84–89).

You can remember the planets' names, in order of distance from the sun, by just the first letters of "My Very Educated Mother Just Sent Us Nachos": Mercury, Venus, Earth, Mars, Jupiter, Saturn, Uranus, Neptune. Pluto is now a dwarf planet, along with Eris, Makemake, Haumea, and the asteroid Ceres (the only object that is both an asteroid and a dwarf planet).

Twinkling Stars and Steady Planets

The stars usually appear to twinkle in the sky. They are tiny points of light far beyond the Earth's atmosphere, and the shimmering and shaking of our atmosphere make the images change in brightness and dance around from moment to moment. The planets are small disks rather than points of light, though we cannot detect this with our unaided eyes. Because planets are disks, their images change and move differently, so the light from a planet appears to be steadier than the light from a star. When you see a celestial object shine steadily while others around it are twinkling, you are probably looking at a planet.

Planets in Our Sky

Venus's orbit around the sun is inside the Earth's orbit, so Venus never appears very far from the sun in the sky. When Venus is visible, it is the brightest object in the sky other than the sun and moon. A bright object shining steadily in the western sky at or after sunset, or in the eastern sky at or before sunrise, is probably Venus.

Jupiter can also be very bright. It can appear in the sky even at midnight. If you see a bright object shining steadily in the southern half of the sky, far from where the sun has set or will rise, you are probably looking at Jupiter.

Mars rarely becomes as bright as Jupiter or Venus, and Saturn never becomes that bright, but they still shine steadily. Mars has a reddish tinge, which is visible even with the naked eye. Saturn appears slightly yellowish. Binoculars are needed to see Uranus or Neptune. It takes a telescope to see Pluto or the other dwarf planets.

Paths of the Planets in Our Sky

The Earth and the other planets orbit the sun in more or less the same flat plane; that is, as though they were all traveling around the surface of a dish. Thus, the sun and the other planets all appear to travel along the same path across the sky. This path, which is traveled precisely by the sun and approximately by the other planets, is called the *ecliptic*.

The ecliptic is shown as a dotted line on the star maps on pp. 20–43. You will always find the planets close to the ecliptic.

The positions of the planets change from week to week and do not repeat from year to year. If a planet is "up," you can find it by looking for a steady object near the ecliptic.

In 1610, Galileo discovered four moons around Jupiter. Here we see space close-ups of their surfaces (left to right in order of their distance from Jupiter and in accurate relative sizes: Io, Europa, Ganymede, and Callisto, to use names assigned later), though in Galileo's small telescope, they were mere points. Our small telescopes, though, can see not only moons around Jupiter, Saturn, and Neptune but also bands in Jupiter's atmosphere and rings around Saturn as well as phases of Venus and Mercury.

HOW TO USE THE STAR MAPS

The pages that follow contain pairs of star maps, each good for certain dates and times. The first map of each pair is for when you are facing north, and the second is for when you are facing south. Together, they show most of the sky, although the very top of the sky, the circle 30° in radius around the zeniths, is not shown. The bright star Vega, for example, is near the zenith on summer evenings, though it is too high to appear on Maps 7 and 8 (facing north).

The dates and times to use each pair of maps are shown along the bottom. The next pair of maps shows the sky either one month later at the same time or on the same date two hours later.

The maps show all stars brighter than magnitude 4.5 and a few fainter ones; you should be able to see all of these stars if you are in a dark place far from cities and with a clear horizon. The area closest to the horizon—the bottom 10° or more, equivalent to the width of your fist at the end of your outstretched arm against the sky or to a thumb's width on the star map—is usually masked by haze or buildings and is therefore hard to see.

Names and colors are shown for all stars that are first magnitude or brighter. The colors of stars reveal their temperatures: bluish-white stars (types O and B) are the hottest; yellower stars, like the sun (type G), are cooler; and reddish stars (type M) are the coolest. (See key on p. 19.)

Some stars (such as Mira in the constellation Cetus) vary in brightness and are shown as open circles on the maps. *Double stars*, pairs or a few stars orbiting each other, are marked with a horizontal line through them. In some regions of space, many stars are found close together. These *star clusters* come in two kinds. *Open clusters*, like the Pleiades and the Hyades (see p. 56), are made up of many young stars formed close to one another. *Globular clusters*, marked with dotted circles, are many old stars packed in spherical form; they look like hazy cotton balls when viewed through small telescopes. A *planetary nebula* is a cloud of gas and dust surrounding a dying star, and a *diffuse nebula* is a cloud of gas and dust in space. *Galaxies* are giant systems of billions of stars, gas, and dust. A few of these nonstellar objects are named on the star maps.

You will notice that some star clusters on the maps have names that consist of the letter *M* and a number, such as M13. M13 is the globular cluster in Hercules that can be seen with the naked eye (see p. 60); it is also the thirteenth item on a list compiled more than two hundred years ago by the French astronomer Charles Messier. Messier made a list of about one hundred fuzzy-looking objects that were always in the sky so that he would not confuse them with the comets he was searching for. We now use his list to refer to many interesting objects in the sky; most, however, can be seen only with binoculars or a telescope.

The ecliptic, shown on the maps as a dotted line, marks the path of the sun and the approximate path of the planets. The planets will always be found close to this imaginary line across the sky.

LEGEND-KEY	STELLAR MAGNITUDES:
	☀ ☀ ☀ ● ∙ ∙
	-1 0 1 2 3 4
	◆ ✦ DOUBLE STARS
	◉ ∘ VARIABLE STARS
	◯ OPEN STAR CLUSTER
	⊕ GLOBULAR STAR CLUSTER
	✧ PLANETARY NEBULA
	▢ DIFFUSE NEBULA
	◯ GALAXY
	SPECTRAL TYPE (Brightest stars only)
	☀ ☀ ☀ ☀ ☀ ☀
	O,B A F G K M

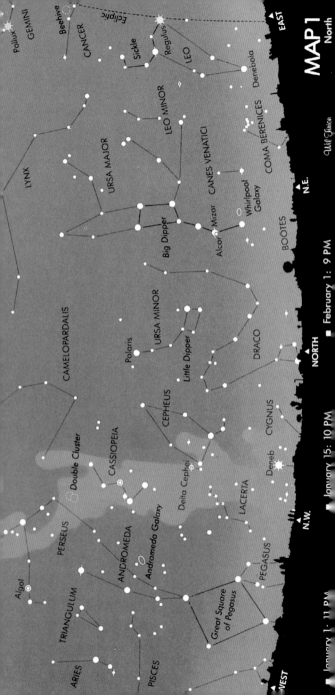

MAP 1
North

Pollux
GEMINI
Beehive
CANCER
Sickle
Regulus
LEO
Denebola
LEO MINOR
CANES VENATICI
COMA BERENICES
LYNX
URSA MAJOR
Whirlpool Galaxy
Mizar
Alcor
Big Dipper
BOOTES
CAMELOPARDALIS
URSA MINOR
Polaris
Little Dipper
DRACO
CEPHEUS
Double Cluster
CASSIOPEIA
Delta Cephei
CYGNUS
Deneb
LACERTA
PERSEUS
Algol
ANDROMEDA
Andromeda Galaxy
PEGASUS
TRIANGULUM
Great Square of Pegasus
ARIES
PISCES

Ecliptic

EAST
N.E.
NORTH
N.W.
WEST

February 1: 9 PM
January 15: 10 PM
January 1: 11 PM

Well Station

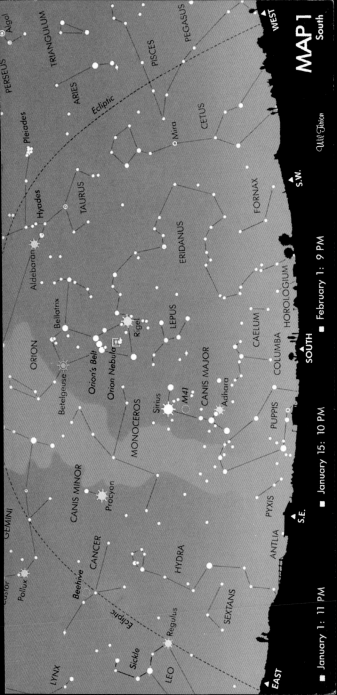

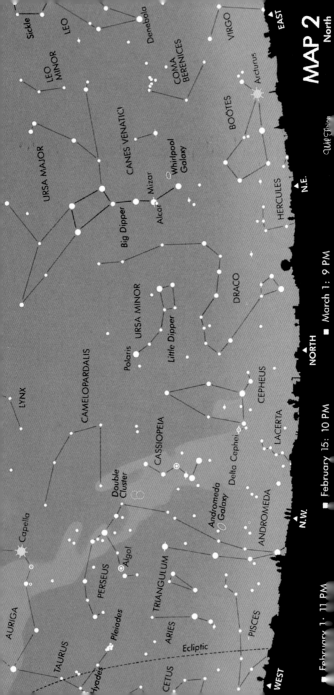

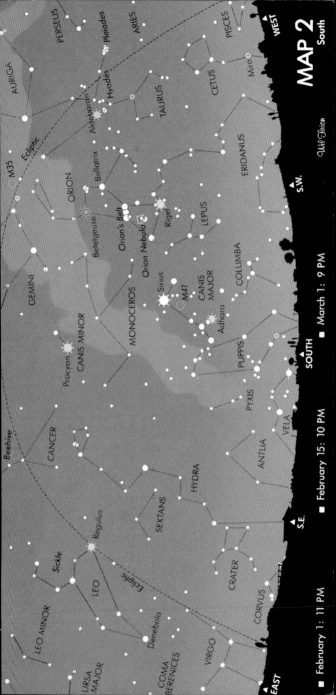

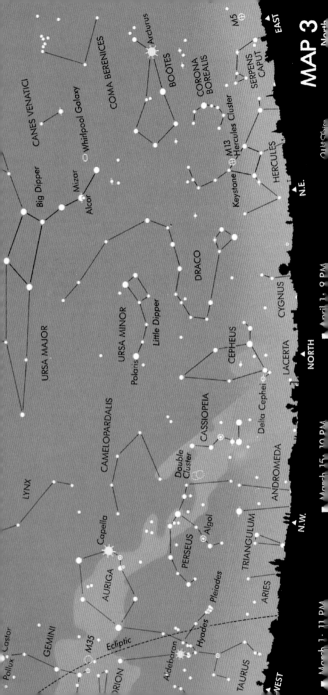

MAP 3

▲ EAST

North

◆ April 1, 9 PM

▲ N.E.

March 15, 10 PM

▲ NORTH

March 1, 11 PM

▲ N.W.

▲ WEST

Constellations and objects labeled:

EAST
M5
SERPENS CAPUT
CORONA BOREALIS
Arcturus
BOOTES
COMA BERENICES
CANES VENATICI
Whirlpool Galaxy
Big Dipper
Mizar
Alcor
URSA MAJOR
M13
Hercules Cluster
Keystone
HERCULES
DRACO
URSA MINOR
Little Dipper
Polaris
CYGNUS
LACERTA
CEPHEUS
Delta Cephei
CASSIOPEIA
CAMELOPARDALIS
Double Cluster
ANDROMEDA
TRIANGULUM
ARIES
Algol
PERSEUS
LYNX
Capella
AURIGA
Pleiades
Hyades
Aldebaran
Ecliptic
TAURUS
M35
GEMINI
Pollux
Castor
ORION

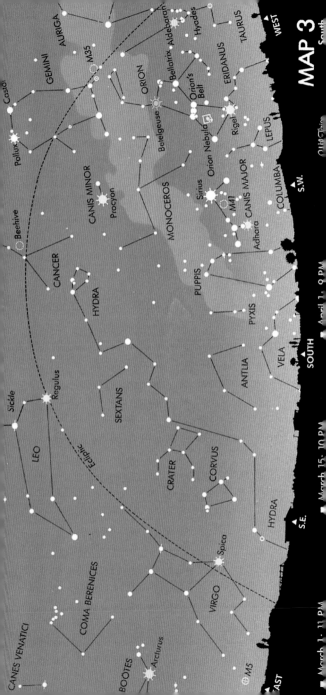

MAP 3

South

WEST

S.W.

SOUTH

S.E.

EAST

■ March 15: 10 PM

▲ April 1: 9 PM

● March 1: 11 PM

Old Chicago

Constellations and stars labeled:

AURIGA
GEMINI
Castor
Pollux
M35
ORION
Bellatrix
Aldebaran
Hyades
TAURUS
Orion's Belt
Betelgeuse
Rigel
ERIDANUS
LEPUS
Orion Nebula
CANIS MINOR
Procyon
MONOCEROS
Sirius
M41
CANIS MAJOR
Adhara
COLUMBA
Beehive
CANCER
HYDRA
PUPPIS
PYXIS
VELA
ANTLIA
Sickle
LEO
Regulus
SEXTANS
Ecliptic
CRATER
CORVUS
HYDRA
CANES VENATICI
COMA BERENICES
Spica
VIRGO
BOÖTES
Arcturus
M5

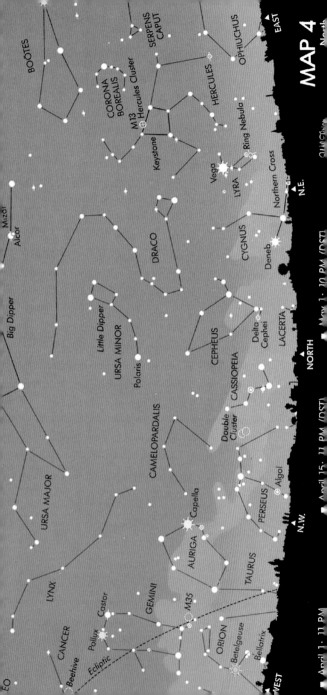

MAP 4

EAST

North

N.E.

NORTH

N.W.

WEST

■ April 1 · 11 PM

■ April 15 · 11 PM (DST)

■ May 1 · 10 PM (DST)

OPHIUCHUS

SERPENS CAPUT

BOOTES

HERCULES

CORONA BOREALIS

M13 Hercules Cluster

Keystone

Vega

LYRA

Ring Nebula

Northern Cross

CYGNUS

Deneb

LACERTA

Delta Cephei

DRACO

Mizar

Alcor

Big Dipper

Little Dipper

URSA MINOR

Polaris

CEPHEUS

CASSIOPEIA

Double Cluster

CAMELOPARDALIS

URSA MAJOR

PERSEUS

Algol

Capella

AURIGA

LYNX

Castor

Pollux

GEMINI

M35

TAURUS

CANCER

Beehive

Ecliptic

ORION

Betelgeuse

Bellatrix

LEO

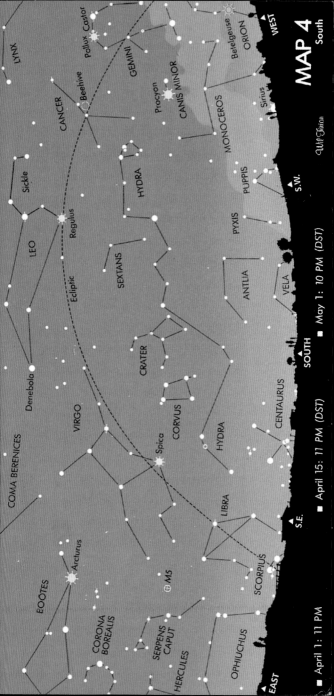

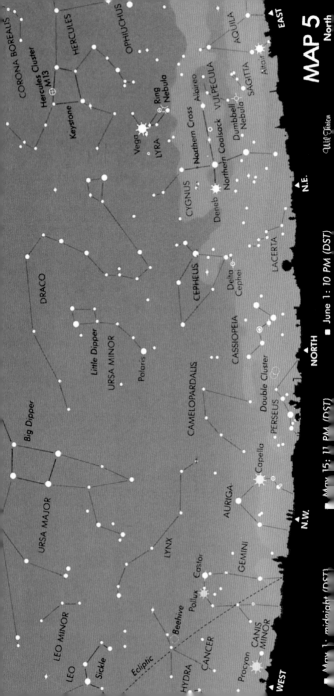

MAP 5
North

CORONA BOREALIS
HERCULES
Hercules Cluster M13
OPHIUCHUS
Keystone
Ring Nebula
Vega
LYRA
CYGNUS
Albireo
Northern Cross
VULPECULA
Northern Coalsack
SAGITTA
AQUILA
EAST
Deneb
Dumbbell Nebula
Altair
DRACO
CEPHEUS
Delta Cephei
LACERTA
N.E.
Little Dipper
URSA MINOR
Polaris
Big Dipper
CAMELOPARDALIS
CASSIOPEIA
Double Cluster
NORTH
URSA MAJOR
PERSEUS
LYNX
Capella
AURIGA
N.W.
LEO MINOR
Castor
Pollux
GEMINI
LEO
Beehive
CANIS MINOR
Sickle
CANCER
Procyon
Ecliptic
HYDRA
WEST

Will Tirion

■ June 1: 10 PM (DST)   ■ May 15: 11 PM (DST)   ■ May 1: midnight (DST)

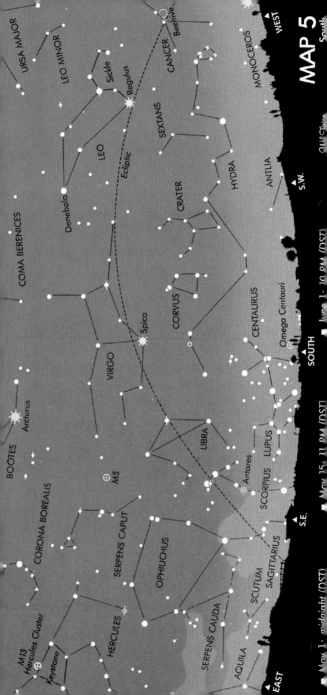

MAP 5

WEST

South

MONOCEROS

S.W.

ANTLIA

Gulf Stars

June 1, 10 PM (DST)

SOUTH

Omega Centauri

CENTAURUS

May 15, 11 PM (DST)

LUPUS

SCORPIUS

Antares

LIBRA

S.E.

URSA MAJOR

LEO MINOR

Sickle

Regulus

CANCER

Beehive

SEXTANS

LEO

Denebola

Ecliptic

HYDRA

CRATER

COMA BERENICES

CORVUS

Spica

VIRGO

BOÖTES

Arcturus

M5

CORONA BOREALIS

SERPENS CAPUT

OPHIUCHUS

M13
Hercules Cluster

Keystone

HERCULES

SCUTUM

SAGITTARIUS

SERPENS CAUDA

AQUILA

EAST

May 1, midnight (DST)

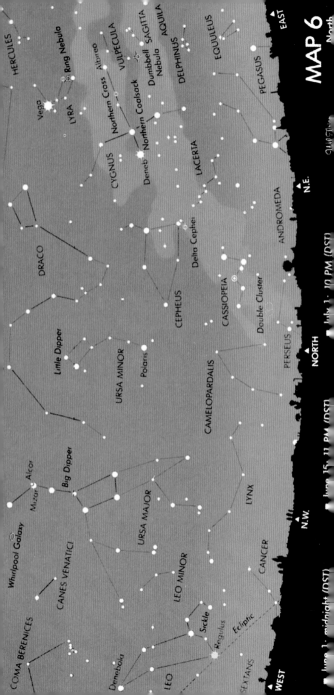

MAP 6

North

WEST ▲
N.W. ▲
NORTH ▲
N.E. ▲
EAST ▲

HERCULES

Ring Nebula

Vega
LYRA

Albireo
Northern Cross
VULPECULA
SAGITTA
AQUILA

CYGNUS

Deneb Northern Coalsack
Dumbbell Nebula

DELPHINUS

EQUULEUS

LACERTA

PEGASUS

Delta Cepheι

DRACO

CEPHEUS

CASSIOPEIA

ANDROMEDA

Little Dipper

URSA MINOR

Polaris

Double Cluster

PERSEUS

CAMELOPARDALIS

Whirlpool Galaxy

Alcor
Mizar
Big Dipper

URSA MAJOR

LYNX

CANES VENATICI

LEO MINOR

COMA BERENICES

CANCER

Sickle
Regulus Ecliptic

LEO

Denebola

SEXTANS

■ June 1; midnight (DST)   ■ June 15; 11 PM (DST)   ■ July 1; 10 PM (DST)

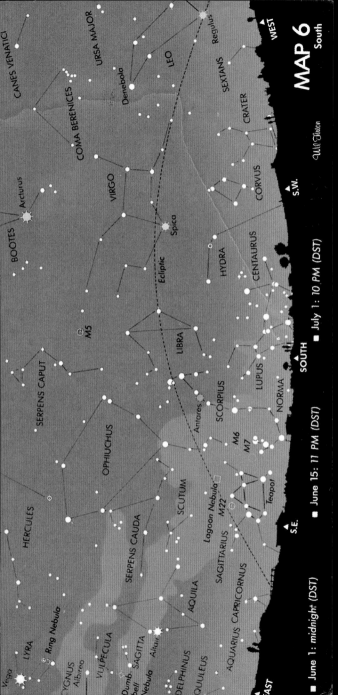

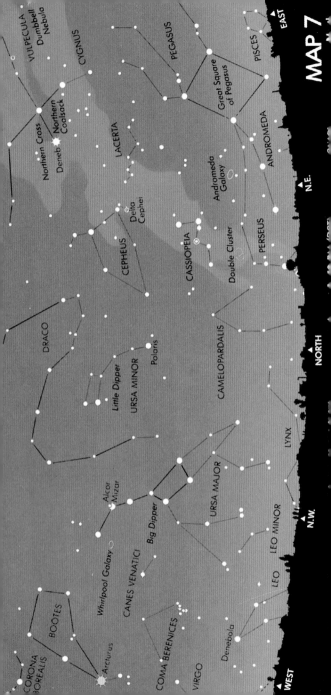

MAP 7

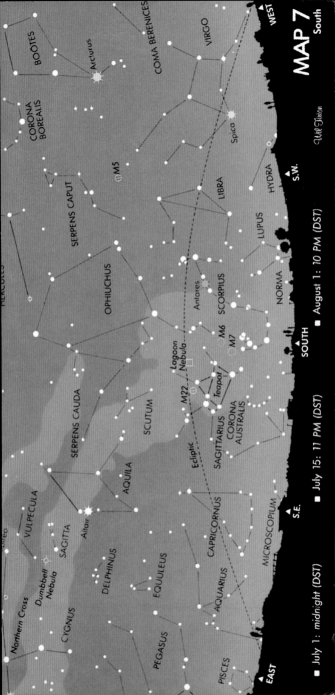

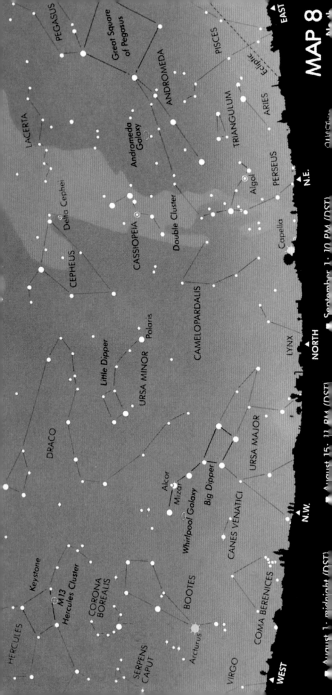

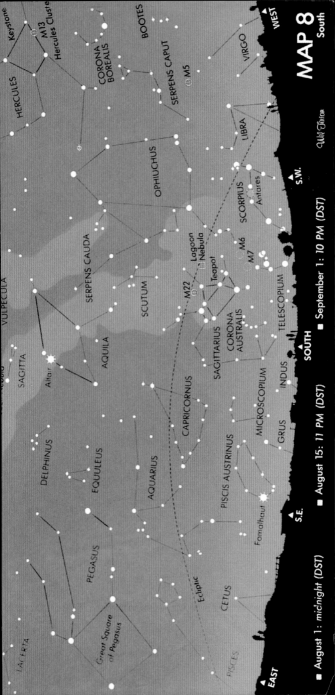

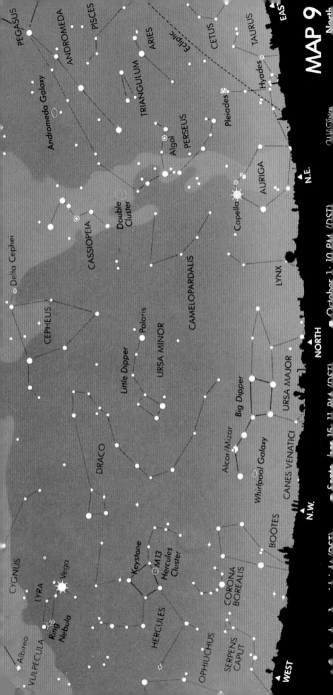

MAP 9

NORTH

EAST

WEST

N.E.

N.W.

North

CYGNUS
LYRA
Vega
Albireo
VULPECULA
Ring Nebula
Keystone
M13 Hercules Cluster
HERCULES
OPHIUCHUS
SERPENS CAPUT
CORONA BOREALIS
BOOTES
CANES VENATICI
Whirlpool Galaxy
Alcor/Mizar
Big Dipper
URSA MAJOR
LYNX
DRACO
CEPHEUS
Delta Cephei
CASSIOPEIA
Little Dipper
Polaris
URSA MINOR
CAMELOPARDALIS
Double Cluster
AURIGA
Capella
PEGASUS
ANDROMEDA
Andromeda Galaxy
PISCES
TRIANGULUM
ARIES
PERSEUS
Algol
CETUS
TAURUS
Hyades
Pleiades
Ecliptic

September 15, 11 PM (DST) · October 1, 10 PM (DST)

MAP 9 South

WEST

SERPENS CAPUT

HERCULES

OPHIUCHUS

LYRA
Vega

CYGNUS
Ring Nebula
Albireo

VULPECULA

SERPENS CAUDA

Lagoon Nebula

Dumbbell Nebula

SAGITTA

Altair

SCUTUM

S.W.

AQUILA

M22

Teapot

DELPHINUS

SAGITTARIUS

EQUULEUS

CAPRICORNUS

MICROSCOPIUM

INDUS

SOUTH

AQUARIUS

PEGASUS

PISCIS AUSTRINUS

GRUS

Fomalhaut

SCULPTOR

Great Square
of Pegasus

Ecliptic

S.E.

ANDROMEDA

PISCES

CETUS

Mira

Andromeda
Galaxy

TRIANGULUM

ARIES

TAURUS

EAST

QuickVision

■ September 1: midnight (DST)   ■ September 15: 11 PM (DST)   ■ October 1: 10 PM (DST)

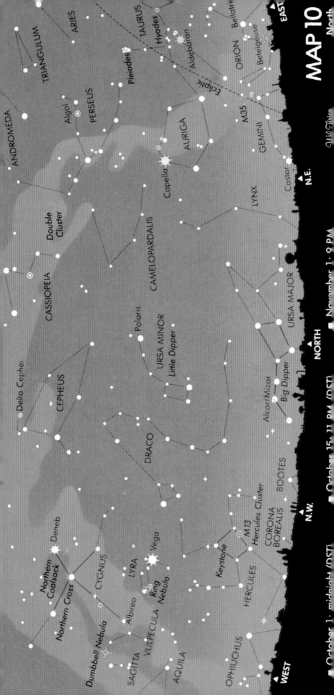

MAP 10

EAST

North

TAURUS
ARIES
TRIANGULUM
Hyades
Aldebaran
Bellatrix
ORION
Betelgeuse
Pleiades
ANDROMEDA
Algol
PERSEUS
Ecliptic
AURIGA
M35
GEMINI
Capella
N.E.
Castor
LYNX
Double Cluster
CASSIOPEIA
CAMELOPARDALIS
Delta Cephei
Polaris
URSA MINOR
URSA MAJOR
CEPHEUS
Little Dipper
Big Dipper
NORTH
DRACO
Alcor/Mizar
BOOTES
N.W.
Deneb
Northern Coalsack
CYGNUS
Northern Cross
Albireo
LYRA
Vega
Ring Nebula
Dumbbell Nebula
SAGITTA
VULPECULA
AQUILA
OPHIUCHUS
WEST
HERCULES
Keystone
M13 Hercules Cluster
CORONA BOREALIS

October 1: midnight (DST) • October 15: 11 PM (DST) • November 1: 9 PM

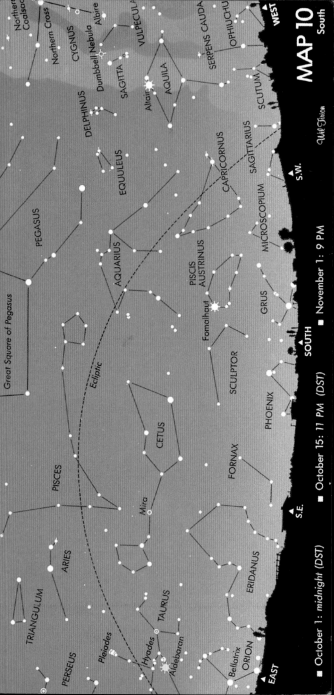

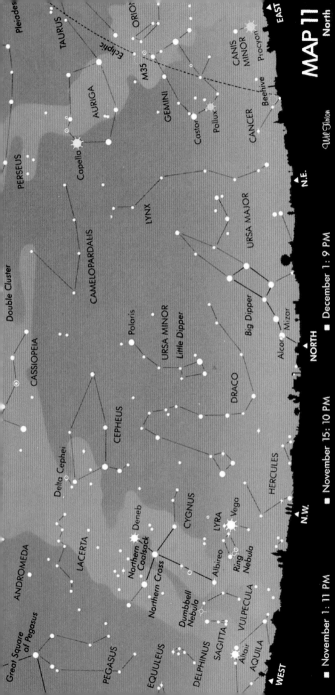

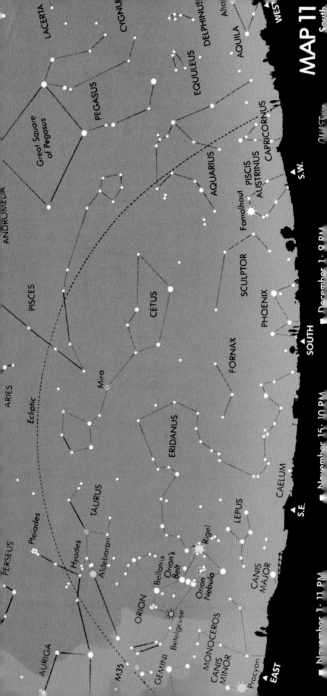

MAP 12

WEST   N.W.   NORTH   N.E.   EAST

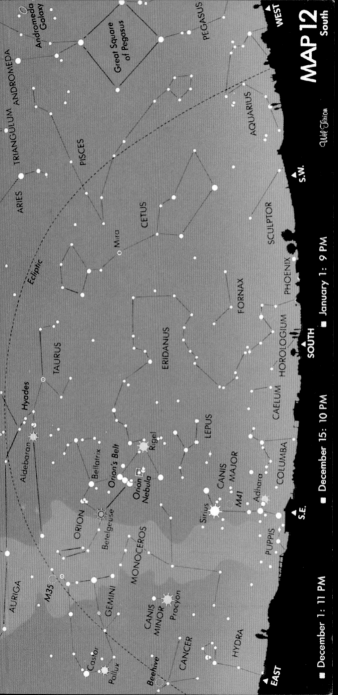

MEET THE CONSTELLATIONS

URSA MAJOR, URSA MINOR

Ursa Major and Ursa Minor, the Big Bear and the Little Bear, are two of the most familiar constellations. For those of us at midnorthern latitudes, they can be seen during all seasons.

In a Greek myth, Zeus, king of the gods, fell in love with Callisto. Together they had a son, Arcas. According to one story, Zeus changed Callisto into a bear to protect her from the jealousy of his wife, Hera. When Arcas grew up, he almost shot his mother by mistake. Zeus turned Arcas into a bear, too, and placed both bears in the sky. He carried them up by their tails, which explains why the bears in the constellation drawings have such long tails. Hera, in her jealousy, convinced the sea god Poseidon not to let the bears bathe in the sea. Indeed, Ursa Major and Ursa Minor are always above the horizon and thus always visible at night.

The Big Dipper and the Little Dipper make up parts of Ursa Major and Ursa Minor, respectively, so they are considered asterisms (see p. 10) rather than constellations. The star in the middle of the Big Dipper's handle is a double star, as you can see with the naked eye if your eyesight is good. The brighter of the two stars is known as Mizar, and the fainter one is Alcor. The Native Americans called them a horse and rider. Seen in a telescope, Mizar turns out to be double itself.

Follow the Pointers of the Big Dipper to Polaris, the North Star, which is at the end of the handle of the Little Dipper. Polaris is the closest bright star to the celestial north pole, only about 1° away from it (twice the diameter of the moon, or half the width of your thumb at the end of your outstretched arm). Polaris orbits the pole once a day. The other stars in the Little Dipper between Polaris and the end of the bowl are hard to see.

Ursa Major is highest in the sky on spring evenings. During the spring and summer, you can follow the arc of the Big Dipper's handle backwards, away from the bowl, to locate the bright star Arcturus.

The constellation Orion, visible on winter evenings, is one of the easiest to find. Three stars of about the same brightness appear in a straight line and form Orion's belt. You can spot them easily when scanning the sky. The reddish star Betelgeuse is about 9° above Orion's belt, and the bluish star Rigel is about an equal distance below it. Betelgeuse is a cool supergiant star, one of the largest stars known. Rigel is a hot star.

In Greek mythology, Orion was a hunter. The god Apollo was afraid that Orion would take advantage of his sister Artemis, the goddess of the hunt. Apollo sent Scorpius, the Scorpion, to attack Orion, who escaped by leaping into the sea. Apollo then tricked his sister into shooting at a dark spot on the waves; her arrow hit Orion, and he died. Artemis could not have Orion revived, but she placed him in the heavens, where the scorpion still pursues him. (See Maps 1–3, 11, and 12, facing south.)

We see in the sky Orion's sword hanging from his belt. A hazy region in the sword, readily visible through binoculars, is the nebula shown below. Deep inside this cloud of glowing gas is a stellar nursery, where stars are being born.

The Orion Nebula, glowing gas in Orion's sword.

CASSIOPEIA

Cassiopeia was a queen of Ethiopia, married to King Cepheus. The boastful queen claimed that she was more beautiful than the sea nymphs. The nymphs had Poseidon, king of the sea, punish Cassiopeia by having the whale Cetus rampage through the kingdom. An oracle said that only the sacrifice to Cetus of Princess Andromeda, daughter of Cepheus and Cassiopeia, would save the kingdom.

All the characters in this tale are visible in the sky. Cassiopeia is marked by its *W* shape; with stars ranging from magnitude 2 to 3.5 marking each turn in the *W*. (Cassiopeia appears on all the northern star maps.)

Cassiopeia lies in the Milky Way, so many nebulae (clouds of gas and dust) and star clusters lie within it. Scanning Cassiopeia with binoculars will reveal many of them.

In the year 1572, a bright new star appeared in Cassiopeia. It was bright enough to shine even during the daytime for a few weeks. This object was a supernova, a star that exploded (see p. 74). Only faint traces of the supernova remain visible today. In the photo below, images of x-rays were used to show the gas thrown out by the supernova, color-coded by x-ray energy.

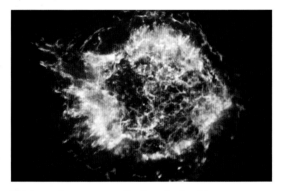

This image shows x-rays coming from the site of the supernova explosion that took place in Cassiopeia in the late 17th century. The remnant's size in the sky is one-quarter the diameter of the moon. The image is from NASA's Chandra X-ray Observatory.

ANDROMEDA

When we left Cassiopeia, the gods had demanded the sacrifice of her daughter Andromeda to save the kingdom. Andromeda was chained to a rock at the edge of the sea, about to be attacked by the whale Cetus. Fortunately, Perseus arrived and turned Cetus into stone by showing him the head of Medusa, a monster he had just killed (see p. 52).

In the sky, the constellation Andromeda is next to Cassiopeia (see Maps 1–3 and 7–10, facing north; it is too near the zenith to show on Maps 11 and 12). Perseus is next to Andromeda.

In the middle of Andromeda you can see a faint, hazy patch of light if the night is very dark. This patch is the Great Galaxy of Andromeda, also known as M31, from its number in Messier's catalog, the 18th-century list of nonstellar objects. It is a spiral galaxy of a trillion stars plus gas and dust; our own Milky Way Galaxy is thought to look much like it. The Andromeda Galaxy is over two million light-years away; that means that the light we see from it today has been traveling for over two million years. This galaxy and one other are the most distant things in the universe that you can see with the naked eye.

The Great Galaxy of Andromeda, M31, a spiral galaxy. Its two companion galaxies are also clearly visible. All three are behind the many foreground stars in our own galaxy.

PERSEUS

Perseus, a hero in Greek mythology, was sent off to kill the sisters known as the Gorgons, winged monsters so horrible that all who looked at them turned into stone. He was helped by the goddess Athena, who gave him a shield so polished that he could see reflections in it. Perseus managed to cut off the head of Medusa, one of the Gorgons, without looking directly at her; he used Medusa's reflection in Athena's shield to aim his sword, so he was not turned into stone. Perseus saved Andromeda, chained to a rock, by showing Medusa's head to the whale Cetus and turning him into stone (see p. 50).

In the sky, Perseus is at the side of Andromeda along the Milky Way. They are high in the sky on a winter evening. If you use binoculars, you can see a pair of open star clusters close together in the midst of the Milky Way in Perseus. This double cluster is shown below.

Another remarkable object in Perseus is the variable star Algol, which is actually a double star. Every 2.9 days, Algol becomes fainter than usual for a period of five hours. It goes from its normal magnitude of 2.2 down to magnitude 3.5 and then back to magnitude 2.2 again. This variation occurs when the dimmer star in the Algol double-star system goes in front of the brighter one.

The double cluster in Perseus, a pair of open clusters known as h and chi Persei. The stars in these clusters are very young.

PEGASUS

When Perseus killed Medusa with his sword, the winged
horse Pegasus arose from Medusa's blood. Pegasus was
wild, and the hero Bellerophon had to tame him. Bellerophon had slain many monsters, and his successes
had made him so vain that he tried to ride Pegasus up
to the gods' home in Olympus. The gods were angered
by his presumption, and Zeus, the king of the gods, sent
a gadfly to sting Pegasus. The horse threw Bellerophon;
he was severely injured, but Pegasus continued on, up to
Mount Olympus and from there to the stars.

In the sky, Pegasus is marked by the Great Square
of Pegasus—four stars in the horse's body that form a
square, each side of which is over 10° (the width of one
fist held up to the sky) across. You can find the Great
Square of Pegasus by following the line from the Pointers of the Big Dipper through Polaris and then twice as
far on Polaris's other side. One of the stars is actually now
over the boundary into the constellation Andromeda—
see Maps 1, 7, and 8 (facing north); 10 (facing south);
11; and 12. The other three stars are in Pegasus proper.
All are between second and third magnitude. There are
no brighter stars in the square.

*M15, a globular cluster in Pegasus, is visible only with binocu-
lars or with a telescope. It is the fifteenth object on Messier's list
(described on p. 19). This is a Hubble Space Telescope view.*

In Greek mythology, Taurus, the Bull, was Zeus himself. He disguised himself as a white bull in order to draw the attention of Europa, princess of Phoenicia. Europa found the bull so attractive that she climbed on his back, and Zeus jumped into the sea and swam off with her.

In the sky, Taurus is charging at Orion (see Maps 1–3 and 10–12, facing south). Only the front part of a bull appears in the constellation. The head of Taurus is clearly marked by the V shape of the open star cluster known as the Hyades. The bright reddish star Aldebaran marks one of the bull's eyes.

The Pleiades, another open cluster, ride on the bull's back. With a little training, you can readily pick up the Pleiades as you scan the sky. The Pleiades are called the Seven Sisters in Greek mythology. Six of the Pleiades' stars are about fourth magnitude and so are easy to see with the naked eye, but some people can see at least two other, fainter stars.

Binoculars or a telescope will reveal dozens of Pleiades stars, as the picture below shows.

The Pleiades: About six of them are visible to the naked eye. Atlas and, above it, Pleione are the two brightest stars at left. Alcyone is in the center of the picture. Continuing counterclockwise around the diamond are Merope, Electra, and Maia. The stars are young and are moving through a dust cloud that is reflecting the stars' bluish light toward us. The Hubble Space Telescope refined our measurement of the distance to the Pleiades as 440 light-years. (Subaru is the Japanese name for the Pleiades, which is why you see a configuration of six stars in the logo for Subaru cars.)

CYGNUS

Cygnus was a friend of the youth Phaethon. Though Phaethon was mortal, his father was Helios, the sun god, who supposedly carried the sun across the sky each day in his chariot. Phaethon begged his father to let him drive the chariot, and one day his father allowed it. But Phaethon drove recklessly and lost control. Zeus had to save the Earth from the sun's heat by hurling a thunderbolt at the chariot. Phaethon fell into a river, and Cygnus, his friend, plunged in to search for him. Helios changed Cygnus into a swan and placed him in the sky.

In the sky, the constellation Cygnus lies along the Milky Way (Maps 4–7 and 9–12, facing north; it is too high to appear on northern Map 8). Its brightest stars mark the Northern Cross. Cygnus, with the bright star Deneb in the swan's tail, appears high in the summer sky. The three bright stars Deneb, Vega (in the constellation Lyra), and Altair (in the constellation Aquila) mark the Summer Triangle. Altair is about 40° from Deneb and Vega.

Albireo, the bright star at the head of the Northern Cross and of Cygnus, is a very pretty double star. With binoculars, you can see that it consists of two stars of different colors.

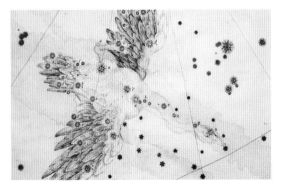

Cygnus, from Bayer's atlas, including the stars that form the Northern Cross.

HERCULES

Hercules was the most famous of the Greek heroes. In mythology, he was the son of Zeus, king of the gods, and Perseus's granddaughter. Zeus's ever-jealous wife, Hera, prevented Hercules from assuming his rightful place on the throne, and even sent two serpents to kill him in his crib. The infant Hercules strangled the serpents. Later, he was forced to serve another king, who ordered him to perform twelve great labors. In the first of the labors, he slew a lion (see Leo, p. 62), and wore its skin thereafter. His other labors included cleaning the stables of King Augeas in a single day, bringing back golden apples from a garden guarded by a dragon, and bringing the three-headed dog Cerberus up from guarding the gates of the underworld. Hercules died on a funeral pyre; the fire consumed his body and sent his spirit to Olympus.

Four stars in the middle of the constellation Hercules form an asterism known as the Keystone (see Maps 3–5 and 8–10, facing north). In the middle of one of its sides, a globular-type star cluster (see p. 73), known as M13 or the Globular Cluster in Hercules, is found. To the naked eye, it is barely visible as a faint patch of light; seen through binoculars or a small telescope, it looks like a cotton ball. The cluster contains about 100,000 stars packed into a relatively small space.

The most prominent globular cluster in the northern sky, M13, is in Hercules.

LEO

In Greek mythology, Leo was the ferocious lion that Hercules slew in the first of his famous labors (see p. 60).

In the sky, you can find Leo by following the back two stars of the Big Dipper's bowl in the opposite direction from that of the Pointers to Polaris. About twice as far along as Polaris is from the Pointers, you come to the bright star Regulus. Regulus marks the dot at the bottom of the backward question mark known as the Sickle. The rest of Leo spreads out behind the Sickle (see Maps 1, 2; 3 and 4, facing south; 5 and 6; 7 and 12, facing north). Leo is highest in the sky at sunset during the winter and spring.

Each year around November 17, the Earth passes through the path of an old comet. As the comet dust burns up in the Earth's atmosphere, we see the Leonid meteor shower, meteors that appear to come from the constellation Leo. The dust is bunched in space, and every 33 years we see a spectacular Leonid shower. The last such spectacular shower, shown below, took place in 1999–2001.

Meteors of the every-33-year exceptionally strong meteor shower, a meteor storm, appear to come from the constellation Leo as they streak across the sky during this time exposure. (The bright stars in Leo's head form the shape of, famously, a sickle.) The next Leonid meteor storm should arrive in 2032.

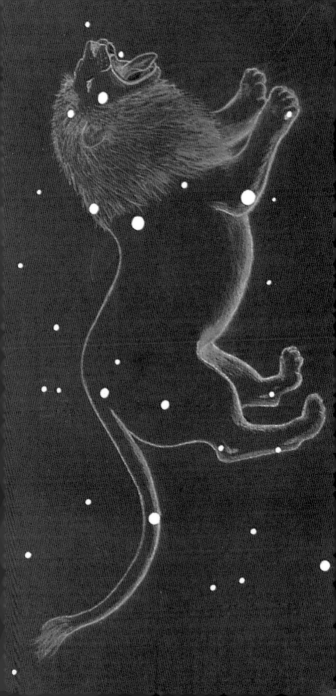

CANIS MAJOR

Canis Major, the Big Dog, is often identified as Orion's companion. In the sky, it is next to Orion (see Maps 1–3 and 11–12, facing south). Canis Major is ready to pounce on Lepus, the Rabbit, one of the animals that Orion hunted.

Sirius, the Dog Star, is the brightest star in Canis Major. It is easy to find, both because it is the brightest star in the sky and because Orion's belt points to it. If Orion is upright, with Betelgeuse at the top, follow the belt to your left.

Sirius becomes visible for the first time each year in August, well after midnight, before the sun brightens the morning sky. This yearly event has given the sultry "dog days of summer" their name.

About 4° (two fingers' width) south of Sirius, in Canis Major, the open star cluster known as M41 (the 41st object in Messier's catalog) appears. Like all open clusters, it contains a few hundred young stars and has no particular shape. Together, the stars are just bright enough to be seen with the naked eye. Viewed with binoculars or a telescope, they make a beautiful sight, covering about as much sky as the full moon does.

Sirius is significant for astronomers because around it orbits a much smaller star, known as Sirius B. Sirius B was the first *white dwarf* star (see p. 74) to be discovered.

M41, an open cluster in Canis Major.

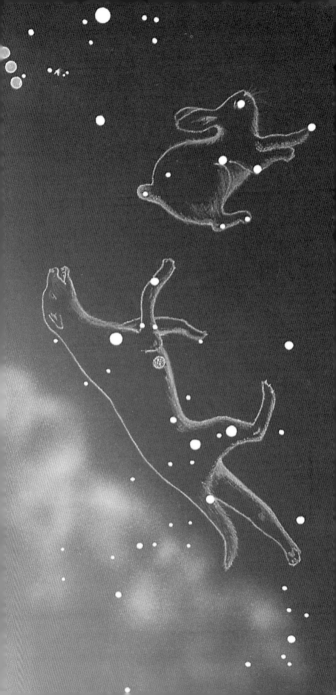

Sagittarius, in Greek mythology, is not only an archer but also a centaur, half human and half horse. Sometimes Sagittarius is identified as Chiron, the wisest centaur.

In the sky, Sagittarius (Maps 5–10) is marked by an asterism known as the Teapot, found near the archer's bow and arrow. Sagittarius appears best on a summer evening, though it never becomes high in the sky for us in midnorthern latitudes. Both the ecliptic—the path where you will find the planets—and the Milky Way go through it. The center of our galaxy is in the direction of the constellation Sagittarius, so the densest part of the Milky Way is there. Thus, in Sagittarius, we find many nebulae and star clusters. In the Lagoon Nebula shown below (M8 in Messier's catalog), glowing gas is given energy by embedded stars. The nebula is crossed by dark dust, which forms the lagoon that can be seen with binoculars or a telescope. The Lagoon Nebula is barely visible to the naked eye as a hazy white region about as large as the full moon.

Within a few degrees of the Lagoon Nebula lie several other beautiful nebulae and star clusters. Scanning this region of the sky with binoculars can be fun.

The Lagoon Nebula in Sagittarius. Dark dust that absorbs light from behind it makes the lagoon that crosses diagonally from upper left to lower right; other dark regions showing absorption by dust are also visible. The nebula glows reddish from the radiation by its hydrogen gas. Such colors are not visible in binoculars or small telescopes; you need long exposures with cameras to show them.

When to See Different Constellations

The Earth rotates on its axis once every day, and it revolves around the sun once each year. But how do we measure rotation? Astronomers measure the length of a day in two ways: with respect to the sun and with respect to the stars. Measurements with respect to the stars are called *sidereal* (sye-dee'ree-al).

When the Earth has rotated once with respect to the sun, one ordinary (solar) day has passed. When the Earth has rotated once with respect to the stars, it has also moved ahead $1/365$ of the distance around its orbit (since there are 365 days in a year). The Earth must then rotate a little farther before an observer on Earth sees the sun come back to the same place in the sky. In fact, it must rotate an additional $1/365$ of a day, which is 3 minutes 56 seconds. Thus a solar day is longer than a sidereal day by 3 minutes 56 seconds.

As a result, constellations rise nearly 4 minutes earlier each day. When a month has passed, they rise about two hours earlier (30 days × 4 minutes per day = 120 minutes). This difference is shown on the star maps (pp. 20–43). We thus see different constellations in each season.

If we were at the North Pole, we would always see the northern constellations and never the southern ones. If we were at the equator, we would see all the constellations, both northern and southern, but each constellation would be up for only half the time. We are at latitudes approximately midway between the equator and the North Pole, so some constellations—the *circumpolar* ones—are always in view, while others are "up" (that is, above the horizon) only part of the time.

Though circumpolar constellations are always in the sky, sometimes they are above the celestial north pole and sometimes they are below it. These constellations will be higher in the sky, and thus usually easier to see, when they are above the celestial north pole.

Circumpolar Constellations (from latitude 40°N)

Ursa Minor
Ursa Major
Camelopardalis
Cassiopeia
Cepheus
Draco

Winter Constellations (Evening Sky)

Northern Sky	Southern Sky
Pegasus	Cetus
Lacerta	Taurus
Andromeda	Orion
Pisces	Eridanus
Triangulum	Lepus
Aries	Canis Major
Perseus	Monoceros
Auriga	Canis Minor
Gemini	Gemini
Cancer	Hydra

Spring Constellations (Evening Sky)

Northern Sky	Southern Sky
Auriga	Virgo
Gemini	Corvus
Lynx	Crater
Canes Venatici	Leo
Boötes	Hydra
Corona Borealis	Cancer
	Gemini

Summer Constellations (Evening Sky)

Northern Sky	Southern Sky
Coma Berenices	Equuleus
Canes Venatici	Delphinus
Boötes	Aquila
Corona Borealis	Scutum
Hercules	Ophiuchus
Lyra	Serpens Caput
Vulpecula	Virgo
Cygnus	Libra
Lacerta	Scorpius

Autumn Constellations (Evening Sky)

Northern Sky	Southern Sky
Corona Borealis	Pisces
Hercules	Aquarius
Lyra	Equuleus
Cygnus	Capricornus
Lacerta	Sagittarius
Pegasus	Ophiuchus
Andromeda	Scutum

LIFE CYCLES OF THE STARS

Nebulae and Star Birth

In between the stars are regions of gas and dust known as *nebulae* (neb'yoo-lee; singular form: *nebula*, neb'yoo-luh). Many of these are visible through binoculars, though they show up only as faint hazy regions, looking almost like clouds. The colors, like those in the photograph below, are too faint for the eye to see. Only long exposures will bring out the colors. Still, it is fun to search the sky to see these hazy shapes.

There are several types of nebulae. Some, like the Horsehead Nebula shown below, contain both glowing gas and dark, light-absorbing dust. The dark dust prevents us from seeing farther. Many of the stars in the image are young stars, formed from the gas and dust in the nebula.

The Pleiades (p. 56) are a cluster of stars embedded in dust. Some of the light from the stars in the Pleiades is reflected to us by the dust. The result is a *reflection nebula*.

Glowing gas, light-absorbing dust, and reflection nebulae are kinds of *diffuse nebulae*.

The Horsehead Nebula, which on the sky is found below the leftmost star of Orion's belt. Dark dust absorbs light from behind it to make the horsehead.

Another type of nebula is a planetary nebula, though it has nothing to do with planets. It got its name because, when seen through telescopes, some planetary nebulae look like small greenish disks, as do the planets Uranus and Neptune. The Ring Nebula, shown below, is one of the easiest to see with a small telescope. It looks like a faint hazy smoke ring; only a long photographic exposure reveals its colors.

Planetary nebulae form when stars like the sun reach the end of their normal lifetimes (10 billion years or so). The sun is about halfway through its lifetime, so we think it will become a planetary nebula in about 5 billion years. First the sun will swell to become a huge, cool star known as a red giant. Then the outer layers of the red giant will drift off into space and become a ring or shell. The hot, lower layers of the star will then become visible as the bluish central star of the planetary nebula.

Stars more massive than the sun will come to a different end: they will explode as supernovae. Their remnants may spread into space and become diffuse nebulae. The Crab Nebula in Taurus and the Veil Nebula in Cygnus are examples of the remains of exploded stars of this type.

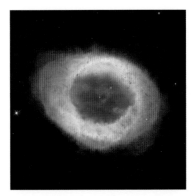

This Hubble Space Telescope view of the Ring Nebula (M57) shows a doughnut shape glowing red from its nitrogen and sulfur gases, pierced by helium gas's blue radiation. In between, hydrogen and oxygen glow in a lighter blue. The Ring Nebula is about 1 light-year across and is 2,000 light-years away.

STARS AND CLUSTERS

The stars are all balls of gas held together by their own gravity. Our sun is an average star. It is about 15,000,000°C (27,000,000°F) at its center and about 6,000°C (11,000°F) at its surface. One million Earths could fit inside it. Some stars are about 15 times less massive than the sun, while others are 60 times more massive. Stars can be about 3 times cooler than the sun and about 10 times hotter.

When you heat an iron poker in a fire, first it glows dimly red, and then brighter red-hot. If it could be made still hotter, it would become blue-white. Similarly, the coolest stars are the reddest ones. Moderately hot stars, like the sun, are yellowish. The hottest stars are blue-white. In the constellation Orion, for example, Betelgeuse (in the shoulder) is a cool, reddish star. Rigel (in the heel) and the stars in the belt are blue-white, hot stars (see p. 11). Astronomers describe the temperatures of stars' surfaces with a series of letters: O, B, A, F, G, K, M, where O stars are the hottest and M stars are the coolest. The sun is a G star. New capabilities that allow us to observe in the infrared have revealed stars even cooler than M, so new spectral types have been added: L, T, and Y (using letters that hadn't already been allocated).

The Beehive Cluster (M44), an open cluster of stars in Cancer. Careful work is often needed to separate the stars in an open cluster from foreground and background stars.

Though the sun is a single star, most stars have close companions. These double stars have two or more stars orbiting around each other, bound together by their mutual gravity. In most cases, it takes years or centuries for the stars to complete one orbit around each other.

Many stars appear in *star clusters*, which come in two basic types. *Open clusters* contain hundreds of stars and have no particular shape. We merely see stars especially close together in the sky. These stars have all formed out of the same interstellar gas and dust. The stars are still young, perhaps "only" hundreds of thousands or millions of years old. The Pleiades in Taurus (p. 56), the Hyades in Taurus, and the Beehive Cluster in Cancer are open clusters that are easy to see with the naked eye or binoculars.

Globular clusters are groups of hundreds of thousands of stars held together in a compact ball. The stars orbit the center of the cluster. They are all very old, about 12 billion years. M13, the thirteenth object on Messier's list (see p. 60), in Hercules, is the easiest globular cluster to see in the northern sky. The bigger the telescope you use, the better you can see the individual stars in a globular cluster.

Omega Centauri, a globular cluster in Centaurus, in the southern sky.

When a star—a hot ball of gas—uses up the nuclear fuel near its center, it collapses. How it ends up depends on how much mass it had. A star with as much or less mass than the sun has gives off a planetary nebula. The nebula's central star cools to become a white dwarf, a tiny star containing perhaps as much mass as the sun but compressed into a size no larger than the Earth. Each teaspoonful of a white dwarf contains many tons of matter. No white dwarfs can be seen with the naked eye. The easiest to detect is Sirius B, a companion to Sirius, though it takes a medium-size telescope to do so.

When a white dwarf is in a double-star system, it can gain some mass from its companion. When a little mass hits the white dwarf and undergoes fusion, the star brightens. We see such a brightening as a *nova*. A nova is visible to the naked eye every few years. Sometimes the white dwarf gains so much mass that it collapses again. It is incinerated, and the star brightens tremendously. Such an event is one kind of *supernova*.

The other kind of supernova occurs when a star more massive than the sun transforms all the hydrogen in its core to iron. Then the star collapses, and explodes.

Supernovae of either kind are very rare. A few are detected by astronomers each year, but almost always in distant galaxies. No supernova has been seen in our own Milky Way Galaxy since the year 1604, five years before the telescope was invented! Nobody saw the supernova of 1667, whose radio and x-ray images we now detect (see p. 48). The first time a supernova was seen with the naked eye after 1604 was in 1987. This supernova became brighter than 3rd magnitude. It occurred in the Large Magellanic Cloud (p. 81), a companion galaxy to our own, and was visible only from far southern latitudes. In 2014, a supernova, the exploded-white-dwarf type, appeared in M82 in Ursa Major, only 11 million light-years from us, but it didn't reach naked-eye brightness by a factor of about 50. This supernova, SN 2014J, is of the type that comes from a white dwarf pushed over its mass limit, making it the closest "standard candle" for comparison with very distant galaxies.

SN 2011fe

TOP: *Supernova 2011fe, an explosion of a massive star, found and studied with the Palomar Transient Factory, a coordination of several telescopes of Caltech's Palomar Observatory on Palomar Mountain, California. The supernova is in the Pinwheel Galaxy (M101) in Ursa Major.*

BOTTOM: *The nearest supernova to us and the brightest in centuries, SN 1987A, showed in 1987 in the Large Magellanic Cloud. The Hubble Space Telescope has tracked a tiny ring of material thrown out by the star before it exploded; the ring, over the past 25 years, has brightened in spots. This whole image is too small to be resolved from the ground.*

PULSARS AND BLACK HOLES

A supernova blasts its outer layers into space. Sometimes we can see the remnant centuries later. The Crab Nebula, below, is the remnant of a supernova that went off over 900 years ago. Though it was the first object Messier listed in his catalog, it is not especially interesting to look at with a small telescope.

The gas in the midst of many supernovae is pressed together during the explosion. It collapses until it becomes a tiny *neutron star*, with as much mass or slightly more mass than the sun. All this mass is contained in a space only about 10 kilometers across, the size of a city. We usually cannot see neutron stars directly, for they are so small and faint. But we can detect some of them from the radio waves they send out. Some neutron stars send out beams of radio waves that sweep around the sky as the neutron stars rotate, just as a beam of light from a lighthouse sweeps around the sky. We detect a pulse of radio waves every time one of these beams passes the Earth, and so we call the objects *pulsars*. The pulsars we know of can rotate as often as 642 times each second or as slowly as once every 4 seconds, still very rapid for a full-fledged star.

A handful of pulsars, including the one in the middle

The Crab Nebula (M1), the remnant of a supernova explosion seen on Earth in the year 1054.

of the Crab Nebula, can be detected optically, but only with large telescopes. The pulsar in the Crab Nebula flashes 30 times each second.

Sometimes the core of a supernova has too much mass for it to stop collapsing at the neutron-star stage. The dying star's gravity, pulling its own mass inward, is too strong. The star keeps collapsing forever. When such a star gets smaller than a certain size, it warps the space around it so much that not even its light gets away. We can think of it as having gravity strong enough to pull back its light. We call these objects *black holes*.

Black holes don't give off light, so we cannot see them directly. We can, however, detect x-rays from the gas just outside the black holes. This gas is in orbit around the black holes. Our atmosphere keeps the x-rays from reaching the Earth's surface, but satellites orbiting the Earth have revealed to us a couple of dozen of such probable black holes. The best-known case is in the constellation Cygnus.

To decide that a black hole is present, astronomers look at the double-star systems from which flickering x-rays are received. By studying whether the single star they see is moving to and fro, they can tell if it has an invisible companion. The hot gas orbiting a black hole may look like the bright gas in the painting below.

An artist's conception of gas pulled off a companion star and then orbiting a black hole in a disk, and then being ejected as a wind.

Our sun and all the stars that we can see are members of a much larger group, called the Milky Way Galaxy, which contains about a trillion stars plus much gas and dust in between the stars. Most of the stars of the Milky Way Galaxy are in a giant disk 100,000 light-years across—the distance that light travels in 100,000 years.

The Milky Way that we see stretching across the nighttime sky is the plane of the disk of the Milky Way Galaxy. From our position two-thirds of the way out, we cannot see through the disk all the way to the center of our galaxy because there is too much gas and dust. When we look toward the center of our galaxy, we see somewhat more stars, gas, and dust than when we look in other directions. This difference explains why the Milky Way looks brightest in the constellation Sagittarius, the direction of the center of our galaxy. In that direction, we see the part of the Milky Way Galaxy that is richest in star clusters and nebulae.

Infrared light penetrates the dust clouds that shield the center of the Milky Way from us, so infrared observations have been used to study stars very near it and how long they take to orbit it. These studies have revealed that a giant black hole with over 4 million times our sun's mass is there.

The center part of the spiral galaxy NGC 2841, showing the lanes of dust and gas that form its arms. It is 46 million light years from us, in Ursa Major.

Many galaxies, like the one shown below, look like our own. It is called the Whirlpool Galaxy, and it can be seen with small telescopes a few degrees south of the star at the end of the handle of the Big Dipper. It is 15 million light-years away, relatively close for a galaxy. That is, the light we are seeing now from that galaxy took about 15 million years to reach us, even at the tremendously high speed at which light travels (300,000 kilometers per second = 186,000 miles per second).

The Whirlpool Galaxy, the Great Galaxy in Andromeda (p. 50), and our own Milky Way Galaxy are all *spiral galaxies*, with arms unwinding in spiral form. The Great Galaxy in Andromeda is the closest spiral to our own and thus the easiest to see.

Near the Great Galaxy in Andromeda are two smaller galaxies. These smaller galaxies are *elliptical galaxies.* Though these elliptical galaxies are relatively small, other elliptical galaxies can be quite large. With a medium-size telescope, you can see many galaxies of all kinds.

The Hubble Space Telescope was used to stare at an apparently blank piece of space. After about ten days of observations, its image of a tiny region revealed thousands of galaxies far out in the universe. It is called the Hubble Deep Field. Later, a Hubble Ultra Deep Field was observed.

The Whirlpool Galaxy (M51) in Canes Venatici, with another galaxy, a small one, at the end of one of its spiral arms. Our own Milky Way Galaxy is also a spiral, though its arms are not quite as open as this galaxy's.

Galaxies are so far away that we can never travel far enough in the universe to see them from another view. To learn about galaxies, therefore, we must look at many of them. We see some of them from the side; others are tilted at an angle, so we can see more of their arms. Ordinary spiral galaxies have arms that unwind like those of a pinwheel; barred spiral galaxies have arms that come off the end of a straight bar that extends to either side of the galaxy's center.

Galaxies are the building blocks of the universe. Astronomers are now mapping how the galaxies are distributed in space. Our galaxy, the Great Galaxy in Andromeda, and about two dozen other galaxies make up the Local Group of galaxies, which in turn is part of a cluster of many galaxies. Clusters of galaxies (see p. 83) seem to be linked in giant filaments that stretch across the universe. In among the filaments are giant voids, where few or no galaxies are found.

It is fun to locate a few galaxies with binoculars. With small telescopes, you can find a larger variety of galaxies, though it is a challenging task.

If you travel far enough south, toward the Earth's equator or into the Southern Hemisphere, the pair of galaxies shown on page 81 make a pretty sight to the naked eye or through binoculars. These galaxies are called the Large Magellanic Cloud and the Small Magellanic Cloud, because Magellan's crew saw them when they

This galaxy, NGC 1300, has spiral arms that seem to come off a bar. Our galaxy has a much smaller bar near its center.

sailed around the world in the 16th century. You must be near or below the equator for them to appear high in the sky.

The word *cloud* is apt, for at first each of these galaxies resembles a hazy white cloud in the sky. They also look like a bit of the Milky Way that has become detached.

A laser beam sent upward from one of the European Southern Observatory's 8.2-meter telescopes makes a "laser guide star" on the Earth's ionosphere, allowing deformable mirrors and computer controls to remove the distorting effect of the Earth's atmosphere to improve the images. In the sky, we see the Large Magellanic Cloud to the beam's left and the Small Magellanic Cloud to the beam's right.

THE PAST AND FUTURE
OF THE UNIVERSE

Astronomers can measure how fast objects in space are moving toward or away from us, and they have discovered that all the distant galaxies are moving away from our own. The farther away a galaxy is, the faster it is moving from us. In other words, the universe is expanding. Our view is like that of someone standing on a raisin in raisin bread dough that is rising. No matter which raisin you stand on, all the other raisins are moving away from you. The farther away another raisin is, the faster it is moving, since there is more material expanding between you and it.

There is, however, no center to the expansion of the universe. The universe fills all space, and space expands as the universe expands. The study of the universe and its past and future is called *cosmology*. Cosmology is advanced by theoretical studies and by studies made with many telescopes on the ground and in space.

Astronomers know the universe is expanding and can see what it was like in the past. Studies tracing the galaxies back in time show that about 13.8 billion years ago, all the matter was compressed densely together. Astronomers have concluded that the universe's expansion began with a Big Bang, which also marked the origin of time.

Astronomers have even detected signals in space that started traveling soon after the Big Bang. With the passage of billions of years, the signals have changed so that they are now received as radio waves. The discovery of these radio waves from the early period of the universe is a proof that the universe did begin in a Big Bang.

What will the universe do in the future? Will it expand forever? Will it ever contract? Astronomers have studied this problem for decades, and the results found in the last years of the 20th century were surprising, as we will see on the next page.

Cosmologists have concluded that during the first second of time, the universe rapidly expanded; that is, it inflated. For a long time, it was thought that gravity from the mass of the objects in the universe, and also from unseen matter known as dark matter, would pull back and slow the expansion. But current studies of distant galaxies reveal that the expansion rate of the universe is accelerating. Some unknown "dark energy" is taking over, pushing galaxies apart faster than they would otherwise be. About 5 per cent of the universe is normal matter, about 25 per cent is dark matter, and the remaining 70 per cent or so is dark energy. The latest version of this result was released in 2013 on the basis of observations from the European Space Agency's Planck spacecraft of the cosmic background radiation, set free only 380,000 years after the Big Bang, some 13.8 billion years ago. Polarization of the background radiation has told us about an inflationary expansion during the first fraction of a second of time.

Thousands of galaxies show in this Hubble eXtreme Deep Field (XDF), a tiny bit of the sky at the center of a long-term Hubble Ultra Deep Field that was compiled from ten years of observations with the Hubble Space Telescope.

The Hubble Deep Field is only about a tenth the size of the moon from side to side. It was taken with the Hubble's Wide Field and Planetary Camera 2. A Hubble Deep Field South was taken three years later, and the Hubble Ultra Deep Field was taken six years still later. The Hubble Extreme Deep Field was released in 2012.

PLANETS AND THE SOLAR SYSTEM

Our sun is surrounded by many smaller objects. Eight of these objects are known as planets. Mercury, Venus, Mars, Jupiter, and Saturn were known even to the ancients. Uranus was discovered in 1781, and Neptune in 1846. Other objects—even the sun and the moon—were called planets at one time or another, as were the first-discovered asteroids. Now we distinguish dwarf planets from planets; the former don't have enough mass to clear out the regions in which they orbit. Dwarf planets (so far) are the outer-solar-system objects Pluto, Eris, Makemake, and Haumea (together known as plutoids) and the asteroid belt's Ceres. In the early 1600s, Johannes Kepler figured out that the planets orbit the sun in elliptical paths (they have non-zero eccentricity; pronounced "ek-sen-*tris*-i-ty"), and linked the duration of their orbits to the orbits' sizes.

In addition to the major planets and dwarf planets, hundreds of thousands of asteroids orbit the sun. Most have orbits between the orbits of Mars and Jupiter. Others, however, come inside the Earth's orbit. Every hundred years or so, an asteroid hits Earth and does noticeable damage; every few hundred thousand years, a larger asteroid hits and does major damage.

Scientists have studied all the planets, several comets, and over a dozen asteroids from spacecraft. These views have revealed aspects of these objects, their surfaces, and their atmospheres that we could not find out with Earth-based telescopes. In the pages that follow, we discuss each of the major objects of the solar system in turn and give some hints on observing them.

Exoplanets

Over 1,000 exoplanets (planet-like objects around distant stars) are known, most by their parent stars dimming periodically or moving toward and away from us as the exoplanets orbit. Statistically, perhaps 90 per cent of exoplanet candidates found by the transit method are thought to be actual exoplanets.

Solar System Data

Planet	Radius of Object (km)	Radius of Orbit (million km)	÷ Earth's	Time for Orbit (yr)
Mercury	2,439	58	0.4	0.24
Venus	6,052	108	0.7	0.62
Earth	6,378	150	1.0	1
Mars	3,393	228	1.5	1.9
Jupiter	71,400	778	5.2	11
Saturn	60,000	1,427	9.5	29
Uranus	26,200	2,871	19.2	84
Neptune	24,300	4,497	30.1	165
Dwarf Planet				
Ceres	476	413.6	2.8	4.6
Pluto	1,150–1,200	5,901.9	39.5	248
Haumea	~500–980	6,430.5	43.0	282
Makemake	~730	6,797.1	45.4	306
Eris	1,163	10,179.7	68.0	561

Moons	Radius of Moon (km)	Radius of Orbit (km)
EARTH:		
The Moon	1,738	384,500
MARS:		
Phobos	13 × 10 × 9	9,378
Deimos	8 × 6 × 5	23,459
JUPITER:		
Io	1,815	422,000
Europa	1,559	671,000
Ganymede	2,631	1,070,000
Callisto	2,400	1,885,000
(plus dozens of smaller moons)		
SATURN:		
Mimas	195	185,600
Enceladus	255	238,100
Tethys	525	294,700
Dione	560	377,500
Rhea	765	527,200
Titan	2,575	1,221,600
Iapetus	730	3,560,000
(plus dozens of smaller moons)		

Moons	Radius of Moon (km)	Radius of Orbit (km)
URANUS:		
Miranda	242	129,800
Ariel	580	191,200
Umbriel	595	266,000
Titania	805	435,800
Oberon	775	582,600
(plus at least 23 smaller moons)		
NEPTUNE:		
Triton	1,300	354,000
Nereid	170	5,570,000
(plus 12 smaller moons)		
PLUTO (dwarf planet):		
Charon	605	17,536
Styx	5–12	42,000
Nix	23–63	48,708
Kerberos	6–17	59,000
Hydra	30–83	64,750
HAUMEA (dwarf planet):		
Namaka	~85	25,657
Hi'iaka	~155	49,880
ERIS (dwarf planet):		
Dysnomia	~350	37,350

Mercury, the planet closest to the sun, has a very hot surface. As it rotates slowly, its surface temperature reaches 400°C (800°F). It is so close to the sun in the sky that it is visible in the west for only an hour or so after sunset or in the east for an hour or so before sunrise. At the times it is visible, it appears as a bright, steadily shining object. Many people have never seen Mercury. Often it is too close to the sun to be seen at all.

Our view of Mercury from Earth must always pass through the turbulent air near the horizon. For decades, our best views of Mercury were from the Mariner 10 spacecraft in 1974. Like our own moon, Mercury is an airless body covered with craters. Mercury also shows some lines of cliffs, as though the planet shrank when it cooled soon after its formation.

NASA's Mercury Surface, Space Environment, Geo-chemistry, and Ranging (MESSENGER) spacecraft was launched in 2004 and went into orbit around Mercury in 2011 (see photo below). BepiColombo (Europe/Japan), with its 2016 launch, will reach Mercury in 2023.

About a dozen times a century, Mercury goes in transit across the face of the sun. Mercury is so small and so far away, about two-thirds of the way toward the sun from Earth, that its silhouette appears smaller than many of the sunspots. The next transit of Mercury to be visible from Earth will be on May 9, 2016. It will be visible throughout the Americas, Europe, and Africa.

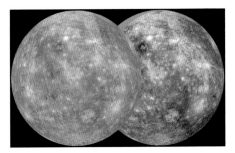

Mercury, imaged from NASA's MESSENGER, with minerals showing at right by enhancing the color contrast.

VENUS

Venus is about the same size as Earth. However, it is closer to the sun, and as a result it is much hotter. Its thick atmosphere traps solar energy inside, making the temperature at Venus's surface 500°C (900°F)—hot enough to melt even lead. The heat is trapped by the *greenhouse effect*: sunlight comes through Venus's clouds and heats Venus's surface, which gives off infrared rays, but the water vapor and carbon dioxide in the atmosphere do not allow the infrared rays to pass through.

Venus can be the brightest object in the sky other than the sun and moon. It can gleam brilliantly in the western sky for two hours or more after sunset or in the eastern sky for two hours or more before sunrise (next page). It is thus often called the evening star or the morning star. Usually Venus shines steadily, but if the air is very turbulent, it can appear to change color, to red and green. A telescope reveals that Venus goes through a set of phases, from crescent-shaped to mostly full.

When Venus passes between us and the sun, it becomes visible against the sun; this event is known as a transit of Venus. Transits come in pairs separated by 8 years; after one pair of transits, the next one does not

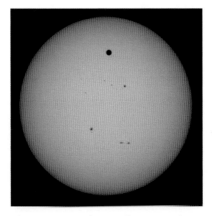

Transits of Venus were seen from Earth in 2004 and 2012, and won't be seen again from our planet until 2117 and 2125. Venus appeared starkly round when sunspots were also present in 2012.

occur for either 105 or 121 years. For twenty minutes or so before Venus entirely enters the sun's disk and similarly after it leaves it, we can see Venus's atmosphere bending sunlight toward us. The transits provide analogs to the transits of exoplanets across their own stars, which we can detect only by those stars' changing light intensities at different wavelengths.

Venus has been visited by many spacecraft from both the United States and the Soviet Union. Radar, which bounces radio waves off an object, penetrates the clouds and has been used to map Venus's surface. Scientists using a small radar on a spacecraft near Venus found that Venus is mostly covered with a vast rolling plain, and has continents. Venus also has volcanoes, which may well be active. The largest is more than 1,500 kilometers across. This volcano, and some other parts of Venus, can also be mapped with powerful radars on Earth.

Spacecraft that have landed on Venus have survived for not more than about two hours because of the high temperatures and pressures. The spacecraft have sent back photographs of flat rocks, like rocks found on Earth, and of an orange sky.

Studying Venus's atmosphere helps us understand our Earth's. We have a greenhouse effect of our own caused largely by our atmosphere's carbon dioxide. Earth's carbon dioxide level, measured in the clean air atop Hawaii's Mauna Loa, a dormant volcano, continues its steady year-by-year increase, with its seasonal fluctuations. Venus's extreme temperature, which results from its greenhouse effect, is a cautionary tale.

Three planets were close together in the sky just after sunset on one evening in 2013. In this image, Jupiter is leftmost; Venus is lowest; and Mercury is to Venus's upper right.

EARTH

The Earth, our own planet, is an oasis in space. The view from space has shown it to be a haven from the harsh conditions that exist at most other places in the solar system.

We now know that the Earth is very active geologically. The continents rest on large flat *plates* that drift slowly around the Earth's surface. Two hundred million years ago, the continents separated from one original land mass or continent. Understanding continental drift, by studying the Earth and comparing it with other planets, can help us understand and predict earthquakes and volcanoes.

Tides in the Earth's oceans are mostly caused by the moon. The moon's gravity pulls water on the near side of the Earth toward the moon and pulls the solid body of the Earth away from water on the Earth's far side. Thus, in most places on Earth, we have two high tides every day.

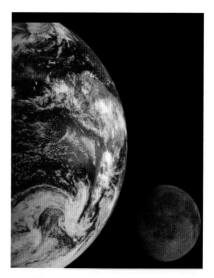

The Earth and the moon imaged from NASA's Galileo spacecraft en route to the Jupiter system. The lunar image shows the rays around the crater Tycho (p. 110).

MARS

Mars appears reddish, even to the naked eye. Since its orbit around the sun lies outside the Earth's orbit, Mars can be opposite the sun in the sky. Thus, it can be up at any time of the night. About every two years, the orbits of Mars and Earth bring them very close together. At those times, Mars appears relatively bright—brighter than the brightest star—and, in a telescope, relatively large. At other times, it can be as faint as second magnitude and relatively small.

In the martian springtime (which occurs at different times in each hemisphere, as on Earth), the surface of the planet changes color. Long ago, it was thought that the color change might be due to growing vegetation, which would indicate that life existed on Mars. We now know that the color change is caused by seasonal winds on Mars, which cover and uncover darker surfaces with reddish dust. The Viking spacecraft that landed on Mars in 1976 found no signs of life, but 2012's Curiosity is finding precursors of life and conditions, such as evidence of flowing water in the past, favorable for life.

Seen in binoculars, Mars looks like a small reddish disk. Even moderate-size telescopes do not show its surface features well.

A mosaic taken with Mastcam on Mars Science Laboratory's Curiosity rover, which landed on Mars in 2012, in the foreground. Mount Sharp, Curiosity's goal, is in the middle background.

In 1976, when the twin Viking spacecraft traveled to Mars, one part of each spacecraft went into orbit around Mars and the other part landed on the planet's surface. The orbiters found giant volcanoes, larger than volcanoes on Earth. They also found vast cratered regions and a canyon as long as the United States measures from coast to coast.

NASA has regularly launched spacecraft to Mars, placing roaming robots aboard Pathfinder (1997; name: Sojourner), Spirit (2004), Opportunity (2004), and Curiosity (2012). The spacecraft have sent back much information about the rocks on the surface of Mars. They have supported indications that water once flowed on Mars's surface, a condition many think is favorable for life to begin.

NASA and the European Space Agency have also sent spacecraft to orbit Mars. The spacecraft take high-resolution photographs and make other observations.

Mars has two small moons, Phobos (from the Greek word for "fear") and Deimos (from the Greek word for "terror"), both named after the mythological companions of the war god, who was called Mars in Rome. These moons are always fainter than 11th magnitude, and cannot be seen from Earth except with large telescopes.

A different panorama from Curiosity, with a billion pixels, made at a site called Rocknest, can be panned and zoomed at the website mars.nasa.gov/bp1/.

Jupiter is the largest of the planets. It is a ball made up mostly of gas, 11 times larger in diameter than the Earth. Its rocky core contains more mass than the entire Earth does.

Jupiter can be very bright in the nighttime sky, brighter than the brightest star. Since its orbit around the sun is outside that of the Earth, it can be on the opposite side of the Earth from the sun. Thus, Jupiter may be seen at midnight or any other hour during the night.

Through binoculars, one can easily see Jupiter's four largest moons: Io, Europa, Ganymede, and Callisto. They were discovered by Galileo in 1609, using one of the first telescopes. Today's binoculars are more powerful than Galileo's telescopes, and the moons can even be imaged with an iPhone. If you watch Jupiter's moons for an hour or more, you can see they are moving with respect to Jupiter; after several hours, you can tell that they are orbiting Jupiter.

Jupiter, a mosaic of images made when NASA's Cassini spacecraft passed by at a distance of only 6.2 million miles (10 million km). We see turbulence in Jupiter's clouds—which are made of ammonia, hydrogen sulfide, and water—and the Great Red Spot, a giant storm.

Even a small telescope shows two or more dark bands on Jupiter. The steadier the air in our atmosphere, the more bands you see.

The Voyager 1 and 2 spacecraft flew by Jupiter in 1979, and the Galileo spacecraft went into orbit in the Jupiter system in 1995. Galileo also dropped a probe into Jupiter's clouds. The Cassini spacecraft also obtained close-ups as it passed by in 2000–2001 en route to Saturn. New Horizons took images in 2007 en route to Pluto. NASA's Juno will arrive in 2016 for a year's orbit. Jupiter Icy Moon Explorer (JUICE) is a planned ESA mission to Ganymede, Callisto, and Europa: proposed launch in 2022 for arrival in 2030.

Jupiter's Great Red Spot, larger than Earth, is a circulating storm that has been present for hundreds of years.

Seen close up, Jupiter's moons were revealed to be worlds unto themselves. They have mountains, craters, valleys, and, in the case of Io, erupting volcanoes.

Jupiter, an image from the Hubble Space Telescope, with a separate image of over a dozen fragments of Comet Shoemaker-Levy 9, which impacted Jupiter and left dark spots for many weeks. We are increasingly aware of objects impacting planets and are increasingly monitoring asteroids and comets to try to detect those that might hit our Earth.

SATURN

Saturn is usually the prettiest object to be seen in the sky, since even a small telescope shows the beauty of its rings. They are chunks of ice and rock orbiting Saturn. Though we now know that Jupiter, Uranus, and Neptune also have rings, only Saturn's are spectacular when seen at a great distance.

Saturn shines as a steady, yellowish light in the nighttime sky and can be up at any hour. A small telescope can show not only the rings but also a dark gap in the rings, called the Cassini division. Our ability to see detail is usually limited by the unsteadiness caused by the Earth's atmosphere rather than by a lack of magnifying power; giving a telescope more magnifying power usually merely enlarges the blurry image but does not make it clearer.

Saturn's moon Titan is one of the largest moons in the solar system. From Earth, it can be seen with binoculars or a small telescope as a point of light near Saturn.

On July 19, 2013, NASA's Cassini spacecraft was eclipsed from sunlight by Saturn, and while in Saturn's shadow, the spacecraft imaged not only Saturn's rings but also a blue dot showing our planet, Earth. The outreach organization Astronomers Without Borders organized a Smile at Saturn activity for us on Earth to participate in while the photograph was being taken.

Since Saturn is farther from the sun than Jupiter and therefore colder, the chemical reactions that create the colors work more slowly in its clouds. As a result, we see fewer contrasting bands on Saturn's surface than we do on Jupiter's. NASA's Voyager spacecraft discovered that Saturn's rings were really made up of many thousands of thin ringlets.

The spacecraft photographed several of Saturn's moons and discovered new ones. Titan has an atmosphere that is even thicker than Earth's, and the surface pressure on Titan is higher than on Earth. Its reddish color comes from a type of smog. Other satellites of Saturn bear cracks, ridges, craters, and many other features. Mimas has a crater so large that the moon must have nearly broken apart when the object that caused the crater hit.

When NASA's Cassini spacecraft reached Saturn in 2004, it went into orbit. It also dropped the European Space Agency's Huygens probe into Titan's smoggy atmosphere. Shorelines and branching riverbeds for liquid hydrocarbons were detected. Enceladus emits plumes of water vapor and ice, revealing a buried ocean that could even contain life.

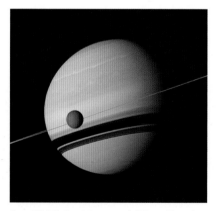

Saturn, from NASA's Cassini spacecraft. We see Titan (3,200 miles across) silhouetted against Saturn's clouds and against Saturn's rings, which are seen nearly edge on. Shadows of the rings appear in the lower half of the image.

URANUS

Uranus, discovered in 1781, was the first planet found that had not been known to the ancients. It is so far away that we knew little about it until Voyager 2 flew close by, in 1986. Until then, Uranus and its moons had been little more than points of light in even a large telescope. In 1977, when Uranus passed in front of a star, the star's light winked off and on a few times before and after the planet hid the star. From this information, astronomers discovered that Uranus was surrounded by nine narrow rings.

The Voyager 2 spacecraft imaged the rings but did not provide much more information than astronomers had from their ground-based studies. Uranus is so far from the sun that its surface is so cold that it is almost featureless. It looks blue-green because methane gas in its clouds absorbs all the other colors. A few individual clouds were barely visible in high-resolution images. By tracing the motion of these clouds, astronomers learned that Uranus rotates on its axis every 17 hours. Modern observing capabilities, with warpable telescope mirrors and the ability to observe in infrared light, have allowed us to study Uranus's weather and surface features from Earth.

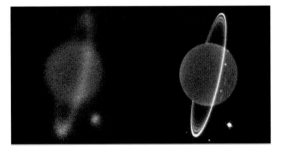

Astronomers are increasingly able to use adaptive optics, with which, controlled by computer feedback, telescope mirrors warp selectively to compensate for the blurring effects of the Earth's atmosphere. Here we see Uranus and its rings and some of its moons with adaptive optics off (left), *and with adaptive optics on* (right).

The moons of Uranus, on the whole, are darker than most of the moons of Saturn, which are icy. Voyager 2 was affected by the gravity of the most massive moons as it passed them, and it could measure how much mass they had. From this information and the moons' sizes, scientists deduced that Uranus's moons are mixtures of rock and ices of water, ammonia, methane, and other chemicals.

The five largest moons of Uranus, those known before the spacecraft visited, are named after characters in Shakespeare's *A Midsummer Night's Dream* and *The Tempest* and in Pope's *The Rape of the Lock*. All Uranus's moons have craters on them. Ariel shows faults on its surface, a sign that geological activity has taken place. Umbriel, about the same size, is completely and uniformly covered with craters. Titania shows few large craters, indicating that material flowed out from its surface to cover them. Oberon has many large craters. Miranda, though smaller, has so many different features on its surface that it obviously underwent some kind of geological activity.

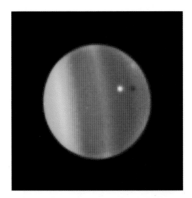

A rare transit of Ariel across Uranus's disk, imaged with the Hubble Space Telescope. The white dot is icy Ariel, one-third the size of our moon, and its shadow appears black. From within the shadow, an observer would experience an eclipse of the sun. The alignment of Uranus's rings with the direction toward the sun that allows such transits occurs only about every dozen years, next in 2019. Even with Hubble's resolution, images of Uranus aren't sharp, as we see in this image.

Neptune, about the same size as Uranus and four times larger in diameter than Earth, remains almost unknown to us. It was discovered, in 1846, as the result of mathematical predictions based on its gravitational effect on Uranus. Like Uranus, Neptune is mostly gas surrounding a liquid ocean and a rocky core.

Neptune has fourteen known moons. The first one discovered, Triton (named after a sea god in Greek mythology who was one of Poseidon's sons), is larger than our own moon. The second moon discovered, Nereid (the Greek word for "sea nymph"), is much smaller and farther away from Neptune than Triton.

From occultations (hidings) of stars as Neptune passed nearby in its slightly different motion in the sky from that of the stars, and by studying the way the stars' images were blocked, astronomers discovered that the planet has rings, though the material in the rings is in clumps rather than continuous.

The Voyager 2 spacecraft reached the Neptune system on August 24, 1989. Neptune then had several giant storms, the largest of which was the Great Dark Spot, which has since disappeared. The surface of Triton, its largest moon, has an icy surface with quite varied terrain and several ice volcanoes. In 2013, with the Hubble Space Telescope, an astronomer discovered a 14th moon.

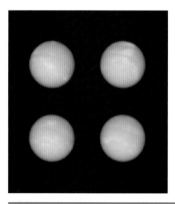

Neptune is only one year old; that is, it has completed only one orbit of the sun between 1846, when it was discovered, and 2011, when the Hubble Space Telescope took this series of birthday photos.

Pluto was discovered in 1930 and was called the ninth planet. In 2015, a NASA spacecraft called New Horizons went up close, revealing detail on it and its moons.

In 1978, an astronomer found a moon of Pluto, which he named Charon. From its orbit, scientists calculated that Pluto's mass is only $1/500$ the mass of our Earth. So Pluto, whose orbit is also very inclined to the orbits of the planets Mercury to Neptune, seemed too small to be a real planet though big enough to be round, and was reclassified in 2006 as a "dwarf planet." (The sun and most stars are "dwarf stars"; there must be hundreds of undiscovered dwarf planets beyond Pluto.) Pluto is the largest dwarf planet; others with names are the asteroid Ceres and objects beyond Neptune called Eris, Makemake, and Haumea.

When Pluto passes in front of a star, blocking it (an "occultation"), the starlight winks out gradually, revealing that Pluto has an atmosphere — since photographed by New Horizons. New Horizons also more accurately measured sizes. Pluto is $2/3$ our moon's diameter and Charon is half that. Pluto's other moons (Nix, Hydra, Kerberos, and Styx) are much smaller.

New Horizons is en route (2019 arrival) to 2014 MU69, one of the thousands of objects orbiting the sun beyond Neptune and Pluto, the Kuiper belt objects.

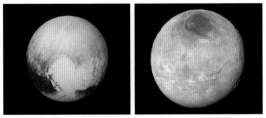

NASA's New Horizons images of (left) Pluto and (right) Charon. The white, heart-shaped feature on Pluto is Tombaugh (after the discoverer of Pluto) Regio.

Pluto's surface is very varied, from smooth regions to mountains as tall as Everest.

Comets

Comets can be especially beautiful to see, but bright ones appear in our sky only rarely. Whenever you hear media reports that a bright comet will be in view, be sure to see it. You may get only a few days' notice.

Comets are chunks of ice that have long been orbiting the sun far beyond Pluto's orbit. The gravity of the Milky Way Galaxy or a passing star sometimes pulls one of these chunks of ice out of its distant orbit. It falls toward the sun. As it nears the sun, the ice turns into gas, and releases the dust in it. Both stream away from the sun in separate tails. The head of the comet contains the icy nucleus with a larger *coma* of gases around it.

Halley's Comet, which last returned in 1986, is the brightest comet that appears at regular intervals. It won't be back near us until 2061. Comets Hale-Bopp in 1997, McNaught in 2009, Lovejoy in 2011, and others appeared brighter to the eye and had longer tails. Sun-observing spacecraft have discovered thousands of comets coming very close to the sun. In 2013, Comet ISON went so close to the sun that it didn't survive.

Comet Hale-Bopp (C/1995 O1) was so bright in 1997 that it could be photographed in seconds with a camera mounted on a tripod.

The advance notice of the return of Halley's Comet meant that many large telescopes could be trained on it and a flotilla of spacecraft from many nations could be sent to visit it, leading to important discoveries. We now know a lot more than we did about the gases in a comet and how they are released.

Photographs of the nucleus of Halley's Comet from the European Space Agency's Giotto spacecraft revealed that the nucleus is potato-shaped and only 15 km × 10 km. The mission confirmed the idea that the comet is a dirty snowball. It is covered with a dark crust that reflects only about 2 per cent of the sunlight that hit it. On the sunlit side, jets of gas and dust come out of a few cracks in the crust. The gas and dust eventually sweep around the nucleus to make the tail.

Several spacecraft have since visited other comets. NASA's Deep Impact dropped a heavy projectile into a comet in 2005 and watched the debris, largely water ice but also some carbon compounds, settle. Europe's Rosetta will land a probe, Philae, on a comet in 2014 and travel for over a year alongside the comet as it approaches the sun.

Comet Lovejoy at Christmas 2011, with a split showing between its curving dust trail and its straight gas tail. The comet's head includes its coma and its nucleus. The moon is at left.

Meteors

As comets orbit the sun, dust spreads out in their orbits. Eventually, the comet itself becomes defunct, either because it uses up most of its ices or because its nucleus is covered with a crust. Still, the dust remains in orbit around the sun. Whenever the Earth passes through the dust, the dust burns up in the Earth's atmosphere. We see these dust grains as shooting stars, called *meteors*.

The Earth travels along the same orbit through the solar system each year, so meteor showers occur annually. The easiest to see is the Perseid meteor shower, so called because the meteors appear to radiate from the constellation Perseus if we trace them back across the sky; however, we can see them all over the sky. The Perseids occur on August 11 or 12 each year, a conveniently warm time to view the skies outdoors.

No equipment is needed to see a meteor shower. Just lie on your back on a lawn chair or a blanket and look up. You may see a meteor every minute or so during the Perseids or the Geminids. Even when a meteor shower is not taking place, you should see a random meteor crossing the sky approximately every 10 minutes. You will see more meteors after midnight than before, since then the Earth has turned so that we are plowing into interplanetary dust.

A wide-field time exposure one minute or so long with an ordinary camera may reveal meteors, especially during a meteor shower. Especially bright meteors, like this one, are called fireballs.

Meteor Showers

Date	Name	Number of Nights Visible
January 3	Quadrantids	1
April 21	Lyrids	2
May 4	Eta Aquarids	3
July 28	Delta Aquarids	7
August 11	Perseids	5
October 21	Orionids	2
November 3	South Taurids	weeks
November 17	Leonids (33-year peak)	weeks
December 13	Geminids	3
December 22	Ursids	2

Asteroids

In addition to the eight major planets and the dwarf planets (and other Kuiper-belt objects), over a hundred thousand minor planets, or *asteroids*, orbit the sun. Most of them are in the asteroid belt between Mars and Jupiter. Many, however, cross the Earth's orbit, and may hit the Earth from time to time.

The largest asteroid, Ceres, is more than 1,000 kilometers across. Vesta, over 500 kilometers across, can become bright enough to be barely visible with the naked eye. You can see more asteroids with binoculars, if you know where to look. NASA's Dawn spacecraft orbited Vesta and moved on to orbit Ceres starting in 2015.

When someone is taking an astronomical photograph using a camera or telescope that tracks the stars, making the stars appear as points of light, asteroids appear as streaks, since they move at a different rate across the sky.

Spacecraft have flown close to several asteroids, sending back pictures of their pockmarked surfaces. One spacecraft even orbited an asteroid and landed on it.

Many scientists are increasingly worried that an asteroid could impact Earth decades or centuries from now, causing much damage. We already know of over 10,000 near-Earth asteroids. NASA's OSIRIS-REx is scheduled to visit a near-Earth asteroid in 2018 and bring a sample back to Earth in 2023.

The moon is the brightest object in the sky aside from the sun. The moon can be seen whenever it is up, even in the daytime sky if it is clear enough.

The moon, which is more than one-fourth the Earth's diameter, orbits the Earth at an average distance of 239,000 miles (384,000 kilometers). With respect to the stars, the moon takes 27 days, 7 hours, and 43 minutes to complete each orbit. The Earth has moved in that time, so it takes 29⅓ days for the moon to come back to the same place in our sky.

In that 29⅓–day period, the moon goes through a whole cycle of phases. The phase we see depends on the relative positions of the Earth, the moon, and the sun. When the moon is on the opposite side of the Earth from the sun, we see its entire sunlit disk. We call this phase a *full moon*. When the moon is one-fourth of the way farther around its orbit, we call the phase the *third-quarter moon*. When the moon is halfway around its orbit, in more or less the same direction as the sun, we have a *new moon*. About a week later, when the moon is another one-fourth of the way around, we have the *first-quarter moon*.

At the first-quarter and third-quarter moons, half the disk of the moon is lighted. When less than half the disk is lighted, we see a *crescent moon*. When more than half is lighted, we see a *gibbous moon*.

During the approximately two weeks from new moon to full moon, we see more and more of the moon each night. The moon is said to be *waxing* (see photos). From full moon to new moon, we see less and less of the moon each night. The moon is said to be *waning*.

On a clear night, you can see the unlighted part of the disk of the moon even when only a crescent is lighted. This "old moon in the new moon's arms" (p. 108) occurs because some light from the sun reflects off the Earth onto all of the moon's disk that is facing us.

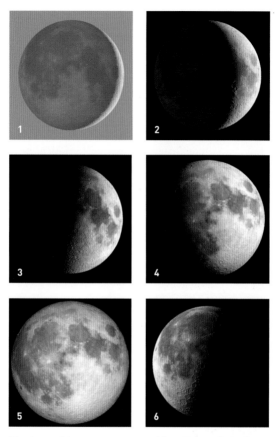

The phases of the moon, reconstructed from high-resolution lunar mapping. We see (left to right, top to bottom): (1) just past new; (2) waxing crescent; (3) first quarter; (4) waxing gibbous; (5) full; (6) third quarter. Because of the moon's elliptical orbit and its inclination, about ⅝ of the moon is visible at one time or another, as we see on p. 107, where features are labeled.

The first-quarter moon shows especially Mare Serenitatis and Mare Tranquillitatis (where Apollo 11 landed), as well as Mare Crisium near the right limb. The third-quarter moon shows especially Oceanus Procellarum. The waxing gibbous and full phase nicely show the crater Tycho, with the rays of lighter material sent out by the impact that made it.

Since the full moon is opposite the sun in the sky, it necessarily rises as the sun sets. Thus, the full moon rises at sunset and is up all night. The moon rises about 50 minutes later each day. About a week later, we have a third-quarter moon, which rises at midnight. The new moon, about a week later, rises around sunrise, but we usually don't notice it since it is up in the daytime. It sets at sunset. A couple of days later, we see a crescent in the west at sunset. After a few more days have passed, the first-quarter moon rises at noon, and so is high above the horizon at sunset.

Galileo, in 1609, discovered craters and flat areas that he called *maria* (pronounced "mar'ee-ya"; the singular, *mare*, which comes from the Latin word meaning "sea," is pronounced "mar'ay"). You can see even with your naked eye that parts of the moon are darker than others. Binoculars or a telescope will reveal the craters and maria.

When trying to use charts showing the moon's maria and craters, remember that though binoculars give an image oriented the same way the image appears to your eye, most telescopes give inverted images. That problem is easy to overcome by turning your moon chart upside down. However, some telescopes have an extra mirror in the eyepiece—a diagonal mirror—that allows you to look into the telescope from the side. Such a mirror flips the image from side to side. Some smartphone apps provide such a flip as an alternative.

One of Galileo's images of the moon, engraved following his drawing and his instructions in his Sidereus Nuncius, *published in 1610. In that book, he reported on his discoveries with the telescope, the first to be turned to the sky for observations and interpreted with the Renaissance era's knowledge of light and shade.*

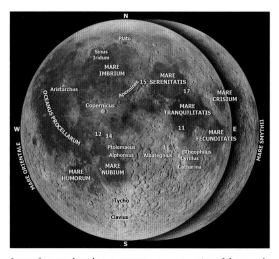

Lunar features, based on a computer reconstruction of the moon's surface using observations from NASA's Lunar Reconnaissance Observer. Views are shown at opposite extremes of the libration, the view over the top or bottom of the moon caused by its tilt and around the sides by its varying speed in its elliptical orbit around the Earth. The librations allow us to view about ⅝ of the moon from Earth, even though the near side of the moon is gravitationally locked toward us.

Some recent space efforts for robotic lunar exploration: the Japanese Kaguya (2007), Chinese Chang'e 1 (2007) and Chang'e 2 (2010), and Indian Chandrayaan-1 (2008) orbited the moon. NASA's LRO (Lunar Reconnaissance Orbiter) reached the moon in 2009, and its LCROSS (Lunar Crater Observation and Sensing Satellite) crashed into a crater that year. NASA's GRAIL (Gravity Recovery and Interior Laboratory) orbited the moon in 2012 to map its gravitational field. NASA's LADEE (Lunar Atmosphere and Dust Environment Explorer) went into lunar orbit in 2013. China's Chang'e 3 mission landed its Yutu (Jade Rabbit) rover on the moon in 2013. Scheduled for 2017: NASA's/Canada's RESOLVE (Regolith and Environment Science and Oxygen and Lunar Volatile Extraction) mission, with a rover to explore water ice near the lunar south pole.

The crescent moon seen in the western sky soon after sunset. Mercury is near it. Earthshine—sunlight reflected off the Earth— allows us to see "the old moon in the new moon's arms."

The moon's orbit is tilted 5° with respect to the Earth's, so usually the moon does not enter the Earth's shadow each month. When it does, we have a lunar eclipse, with the sun, Earth, and moon in a line. The darkest part of the Earth's shadow, the *umbra* of the shadow, takes about 2 hours to cover the moon. The total phase of the lunar eclipse may last an hour or so. Some of the red part of the sun's light is bent through the Earth's atmosphere (the blue part is scattered en route to make blue skies on Earth, so doesn't get through), so the totally eclipsed moon usually appears faintly red. A lunar eclipse can be seen by anyone for whom the moon is up. Total lunar eclipses 2014–2019: April 15, 2014 (Americas—total throughout except for setting while eclipsed in the Northeast; Pacific, Australia); October 8, 2014 (Americas—total in Hawaii, Alaska, and the western U.S., and setting while still partial for the rest of the continental U.S.; Pacific, Australia, Asia); April 4, 2015 (Americas—total in Hawaii and Alaska, and set while still partial for the continental U.S.; Pacific, Australia, Asia), and September 28, 2015 (Americas—total in the eastern U.S. and Canada, and rising while partial farther west; total also in western Europe, western Africa, and western Asia). After that, the next lunar eclipses will not occur until January 31, 2018; July 27, 2018; and January 21, 2019. Maps can be linked from the NASA website eclipse.gsfc.nasa.gov/LEdecade/LEdecade2011.html.

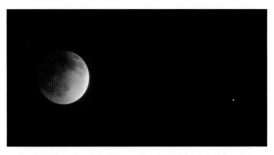

In the top photo, from the lunar eclipse of April 14–15, 2014, the bright star Spica is at the right.

The totally eclipsed moon glows reddish because of light from the sun that has passed through the Earth's atmosphere. Blue sunlight is scattered more by air molecules than red sunlight, so the blue is scattered out to make blue skies on Earth and only the red survives the passage through the Earth's atmosphere to reach the moon during a lunar eclipse. The composite (center above) of the 2011 eclipse shows that the Earth's shadow is round, as was known even at the time of Columbus. (lowest image) We see a composite of images taken during the April 14–15, 2014, total lunar eclipse over the three and a half hour period from the beginning of the partial umbral eclipse to the end of the partial umbral eclipse. In the middle of that period, the moon was totally eclipsed for one hour and 18 minutes. The penumbral eclipse, which is barely visible to the eye, lasted about five hours and 45 minutes.

The crater Tycho (named after Tycho Brahe, the great 16th-century astronomer whose observations enabled Johannes Kepler to find his laws of planetary orbits), observed close up by NASA's Lunar Reconnaissance Observatory, launched in 2009.

TOP: *Tycho's central peak, about 10 miles (15 km) wide.*
BOTTOM: *Tycho's summit.*

The moon was visited by a series of American spacecraft, some manned and some unmanned; the most famous visits were the Apollo missions, which brought twelve astronauts to the moon from 1969 to 1972. The astronauts took many measurements on the moon, left behind reflecting mirrors to bounce back laser beams from Earth, and carried home a total of 842 pounds (382 kilograms) of moon rocks.

The scientists analyzing the data discovered that the craters on the moon were formed by meteorites—rocks from outer space hitting the moon's surface. A few craters, like Copernicus and Tycho (pp. 107, 110), show rays of light-colored material thrown out in the impact.

The rays darken with time, so Copernicus must be one of the younger lunar craters, perhaps only a few hundred million years old. Most of the other lunar features have been in place for billions of years. The highland rocks—the heavily cratered regions—were formed between 4.42 and 3.9 billion years ago. Lava flowed over the lunar surface to form the maria (lunar seas) between 3.8 and 3.1 billion years ago. The few craters in the maria formed after that time.

The last people on the moon: Harrison Schmitt (visible in reflection in Gene Cernan's helmet) took this photo during the Apollo 17 mission in 1972. Both China and India have announced plans to send people to the moon, though the landings probably wouldn't take place before 2020.

THE SUN

The sun is our nearest star. It differs from other astronomical objects mainly in that we can observe it only in the daytime. The sun is so bright—magnitude –27—that its light scatters around in the Earth's atmosphere, making our sky bright. The blue part of the sun's light scatters more efficiently than its red light, so the daytime sky is blue.

The sun is so bright that it can burn your retina—the back part of the inside of your eye—if you stare at it. So never stare at the sun, and never look at the sun through binoculars or a telescope unless special solar filters are in place, or during totality at an eclipse.

You can make a solar image that you can safely look at by projecting it onto a screen. You can use binoculars or a small telescope to do so, adjusting the focus so that the image is clear on the screen when you hold the binoculars or telescope about 3 feet (1 meter) from the screen. You should look *only at the image on the screen*, and never up at the sun through the binoculars or telescope.

When you look at a solar image, you usually see a few dark regions—sunspots—on the sun. Each sunspot is a region that is about 1000°C cooler than the rest of the sun's surface. The sun's magnetic field is concentrated there.

The number of sunspots on the sun rises and falls with a period of about 11 years. The most recent time of sunspot minimum, the part of the cycle when few sunspots were visible, were the years 2008 through 2010. In 2013–2014, we had sunspot maximum, when there were many spots every day, though the peak was not as high as at the last few sunspot maxima. The next sunspot minimum will be in about 2020.

If you become more interested in solar observing, you can purchase special filters that will show more structure on the sun's surface or above the sun's edge, as can be seen on the next page. The number of features that can be seen also varies with the sunspot cycle.

The features we see on the sun presumably exist on other stars as well, though the other stars are too far away for us to see such detail.

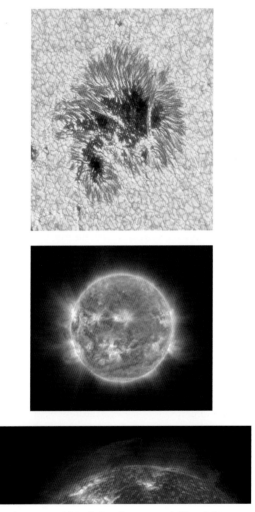

TOP: *A sunspot, with its dark umbra surrounded by a lighter penumbra. Around the sunspot, we see the salt-and-pepper effect called* solar granulation, *an effect like water boiling in a pot on the stove.* MIDDLE: *A view of the sun in ultraviolet radiation from the solar corona, taken with an ESA/Royal Observatory Belgium Sun Watcher camera in space at the same time as the 2012 total solar eclipse (pp. 115 and 116).* BOTTOM: *Solar prominences seen above the sun's limb from NASA's Solar Dynamics Observatory in 2013.*

Because of the tilt of the moon's orbit with respect to the Earth's, the moon does not pass directly between the Earth and sun each month. But about every year and a half, the moon does pass between the Earth and sun, blocking the sunlight from reaching a small region of Earth. As the sun, moon, and Earth move in space, the shadow cast by the moon sweeps a long, thin path across the Earth. Along this path—thousands of kilometers long and as much as hundreds of kilometers wide—we see a *total solar eclipse.*

A total solar eclipse is the most spectacular astronomical sight that exists. Over a period of an hour or two, the moon gradually blocks the sun. The darkening of the sky is noticeable only during the last few minutes of this time. Finally, the sky grows rapidly darker, as less and less of the sun is visible. Then the moon covers the entire bright disk of the sun. As it does so, the sun shines through a few valleys on the edge of the moon. These bright beads of light are Baily's beads. The last Baily's bead before the eclipse is total and the first afterward glow so brightly that they look like a diamond on a ring, creating the *diamond-ring effect* (see p. 116). The total part of the eclipse—*totality*—lasts for a period that is sometimes as short as a few seconds or sometimes as long as 7 minutes. At the end of totality, we briefly see the diamond-ring effect again, marking the end of totality. The bright solar surface comes out from behind the moon, and the sky brightens.

During totality, the sky around the sun is dark enough that we can see stars, even though it is daytime. Near the horizon, however, we can see far enough to get light from regions where the eclipse is not total. Because, as at sunset, we are looking through a lot of air, we see a reddish horizon all around, like a 360° sunset.

Eclipses look so beautiful because of a lucky circumstance—the moon is 400 times closer to us than the sun and also 400 times smaller. Thus the sun and the moon take up almost exactly the same angle in the sky, and the moon can block the sun exactly.

Though a total solar eclipse occurs somewhere on Earth every year and a half, it takes more than three hundred years on average for an eclipse to pass your particular location. So people travel to see the beauty of solar eclipses. The November 12, 2012, total solar eclipse was visible only in Australia and the Pacific; the November 3, 2013, total solar eclipse was visible only in the Atlantic and across Africa.

The next total solar eclipses will be visible in the Arctic (Svalbard and Faroe Islands) on March 20, 2015, and in Indonesia's Sumatra and Sulawesi and in the Pacific on March 9, 2016.

The U.S., except for Hawaii, will see a partial solar eclipse on October 23, 2014. The path of totality of the total solar eclipse of August 21, 2017, will cross the continental U.S. from Oregon to South Carolina. People in the rest of the United States and Canada will see a partial eclipse. It is well worth traveling to be in the band of totality. Link to my website for the Working Group on Solar Eclipses of the International Astronomical Union at the website www.eclipses.info for maps.

Sometimes the moon is relatively far from Earth in its elliptical orbit and does not quite cover the sun during an eclipse, even when it passes centrally across it. An *annulus*—a ring—of sun remains visible, so we have an annular eclipse. There was an annular phase at the very beginning of the path of totality on November 3, 2013.

The total solar eclipse of 2012, viewed from a helicopter above the clouds on the northeast coast of Australia. The shadow of the moon, the umbra, is very visible as the wide V of darkening symmetrically around the solar corona, with the silhouette of the lunar disk barely visible.

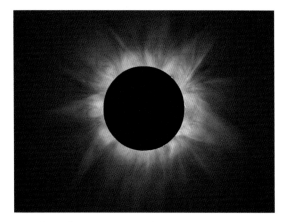

TOP: *A composite image showing the solar corona surrounding pinkish prominences and chromosphere around the silhouette of the moon during the November 14, 2012, total solar eclipse, observed from the Tablelands near the northeast coast of Australia. We see coronal streamers in all directions around the lunar disk, typical of the maximum phase of the sunspot cycle.*

BOTTOM: *A montage showing phases of the annular eclipse of May 10, 2013, observed from the outback in the middle of Australia. The clouds present at the eclipse's beginning (center of the montage) dissipated by the time of annularity.*

You can link to maps of eclipse paths at the websites www.eclipses. info or www.eclipse-maps.com.

Observing Eclipses

The sun is so very bright that it can harm your eyes if you stare at it for more than a few seconds or if you look at it directly through binoculars or a telescope. However, during the total part of an eclipse, the part of the sun you see—the solar corona—is only as bright as the full moon, and so is not hazardous to look at. In an annular eclipse, the solar surface is never fully covered, so do not look directly at it.

To view the partial or annular phases of an eclipse, it is easy to make a pinhole camera. Simply make a hole a few millimeters across in a piece of opaque cardboard, or use an object with small holes, such as a cheese grater. Hold this object up to the sun, and you can make an image of the partially eclipsed sun fall onto a piece of paper or cardboard. Look only at this second object, not up through the hole in the first. For a better image, you can purchase a special solar filter, which is often for sale at a few dollars at eclipse times. It is much, much less transparent than sunglasses, which are not safe to use.

Do travel to the band of totality if at all possible. The path of the August 21, 2017, total solar eclipse, is relatively accessible to Americans.

If you are lucky enough to be in one of the few spots on Earth where the sun is totally eclipsed, make sure you look directly at the sun during the total phase; do not miss the spectacular show. Then turn away when the sun's bright everyday surface comes out again.

LEFT: *Holes punched in paper or ready-made in a cheese grater project pinhole images onto a piece of paper.* RIGHT: *Commercial solar filters reduce the solar intensity to a safe level; some come in eyeglass or viewer-card shapes and are available for a few dollars or less. Coated-glass filters like the one shown here cost a few tens of dollars. See the website www.eclipses.info.*

TIPS ON NIGHTTIME OBSERVING

Observing with Your Eyes Only

The best way to start observing the stars and other bodies in the universe is with your eyes. Even without the aid of a telescope or binoculars, your eyes make fine detectors for light from the stars.

When you are in a bright room, however, the cells in your retina become bleached and your eyes are not very sensitive. When you go out into the night, your pupils widen right away. Each can become 8 millimeters across, instead of the 2 or 3 millimeters across they were inside. But your retinas stay bleached for a while. It may take as long as 15 minutes for your eyes to become *dark adapted*. You will find yourself seeing more and more stars as this time passes.

Be careful not to look at any white lights during this time. Red light will not interfere with your adaptation to darkness, so you can use a red flashlight.

With your naked eyes, you can see the sun, the moon, and stars as faint as 6th magnitude. In particular, you can see all the stars on the star maps in this book. You can see traces of one or two globular clusters, such as M13 in Hercules (p. 60), and the galaxy M31 in Andromeda (p. 50).

You can see five planets with the naked eye: Mercury, Venus, Mars, Jupiter, and Saturn (not to mention the Earth).

You can study certain stars with your naked eye, comparing them with other stars to notice variations in brightness. Every three days, Algol drops a factor of three in brightness over the span of an hour; Betelgeuse changes in brightness over months. You can see Alcor and Mizar as a double star.

If you are lucky, you can see a comet. Comets may stay visible to the naked eye for a few days or a few weeks.

All in all, naked-eye astronomy can be a lot of fun.

Observing with Binoculars

Binoculars gather light and funnel it to your eyes. Thus you can see fainter things with binoculars than you can with your naked eye, and also see more detail. (Never look at the sun directly with binoculars or telescopes, except with special filters or during an eclipse's total phase.)

Binoculars are rated by two numbers: the magnifying power and the diameter of the front lenses. For example, 7 × 50 binoculars magnify seven times and collect light with a pair of lenses that are each 50 millimeters across. Since your pupil opens to 8 millimeters across at best, a binocular lens collects a lot more light than your eye.

For astronomy, we are more concerned with collecting light than we are with magnification. 7 × 50 is the most popular size binocular to use for astronomy. Binoculars with higher magnification cannot be handheld without shaking too much; a tripod helps.

With binoculars, you can see not only the stars on the maps but can also tell that many of them are double. You can see many nebulae, such as the Orion Nebula, and can see star clusters much better. You can see ten or more stars in the open cluster known as the Pleiades instead of the six to eight most people see with the naked eye. If you sweep along the Milky Way, you will find many clusters and nebulae on your own.

Prisms or mirrors inside binoculars reflect the light entering the lenses. Use binoculars without your eyeglasses (if you wear them, and if your astigmatism isn't too bad) and adjust the focus to compensate for your nearsightedness or farsightedness. Binoculars also have an adjustment on one of the eyepieces to allow for a difference between eyes.

Observing with Telescopes

Telescopes collect light and funnel it to your eye or to a photographic plate or electronic detector. They are thus types of light buckets. The magnification effect of telescopes is less important.

Some telescopes collect light using a front lens. These are refracting telescopes. Popular *refracting telescopes* often have lenses that are 2 to 4 inches (5 to 10 cm) across. Some telescopes collect light using a rear mirror. These are *reflecting telescopes*. Popular reflecting telescopes often have mirrors 4 to 8 inches (10 to 20 cm) across. The image quality of a refracting telescope is often equivalent to that of a reflecting telescope about twice as large (an effect most noticeable when you are observing planets).

Compound telescopes that are now very popular have a mirror in the rear but also have a thin lens at the front. The lens allows telescope makers to use a spherical mirror, which is easier to make and allows a wider field of view, rather than a more complicated paraboloidal mirror that can be used without a front lens. The compound telescopes are usually of the Schmidt-Cassegrain type, in which the light that passes through the front lens is reflected up the tube by the spherical main mirror and is then reflected down through a hole in the main mirror, where it is conveniently accessible to an eyepiece or a camera.

Using a refracting telescope in the daytime to project an image of the sun; in such a case, look down at the projected image and never through the telescope at the sun. Here we see the 2012 transit of Venus, with the silhouette of Venus visible against the sun's disk.

The most important thing to look for when you are buying a small telescope is, surprisingly, the quality of the mount. Many people buy a small telescope and then find that the mount that came with it is too shaky to hold the telescope steady. Sturdy mounts are not cheap, but it is a waste of money to buy a telescope with a flimsy mount. A Dobsonian mount, which you point up and down as well as around but that doesn't track the stars, can be made inexpensively of particleboard.

Once you have a telescope, you can collect a lot of light, which will allow you to see much fainter objects than with the naked eye. Also, you can see finer detail, such as the rings of Saturn. You will see more star clusters and more individual stars in them. You will see more types of nebulae and more structure in them. You will see many more galaxies.

Your telescope will probably be on a mount that tracks the stars as they move across the sky. Since the stars appear to move around the Earth once every 24 hours, a simple motor attached to a north pole–pointing axis of a mount, and turning once in 24 hours, allows the telescope to keep up with the stars. Many mounts now have axes that point, instead, only straight up or horizontally, with computers calculating how to keep up with the stars. If you have this type of mount, you can place a camera back (without the lens, because the telescope acts as the lens) on the telescope and take pictures of clusters, galaxies, and nebulae. Or you can attach a camera with its lens on top of the telescope, taking a long exposure while the telescope tracks. Telescopes are now available with GPS and automatic alignment to simplify tracking. Holding a digital camera or smartphone up to the eyepiece of a telescope that is pointing at the moon can produce nice images.

Even without a telescope, you can use a camera to take constellation photos. Just put your camera on a sturdy tripod, raise the camera's speed to perhaps ISO 500 or 1600, and try exposures of 1, 2, 4, 8, and 16 seconds. With an hours-long exposure, you will see star trails (p. 7).

Happy observing.

TIME

Time on our watches is based on the motion of the sun. Basically, when the sun passes due south, it is noon. However, using that system would mean that any person on a different line of longitude would have a different solar time. In 1884, a series of *time zones* was set up; all the people in a certain time zone have the same time. Time jumps, usually by a whole hour, at the edge of a time zone.

When it is six o'clock at night in the summer, it is still light out. To give us more time to spend outdoors in daylight, most states (though not Arizona, except for some Native American reservations, or Hawaii) have adopted daylight saving time, setting the clock forward by one hour in the spring (the second Sunday in March) so that 6:00 p.m. becomes 7:00 p.m. In the fall (the first Sunday in November), we set our clocks back by one hour. Remember: "Spring forward, fall back."

Astronomers often keep time by the sun's position in relation to the zero line of longitude, the meridian of Greenwich, England. They use this *universal time* (UT) so data from all over the world can be easily compared.

Time measured by the stars, *sidereal time* (p. 68), matches solar time at the fall equinox, around September 21. Then sidereal time grows earlier than solar time by 3 minutes 56 seconds each day. At the spring equinox, sidereal time is 12 hours earlier than solar time.

An international organization proposes to change timekeeping—coordinated universal time (UTC)—so that it relies solely on atomic clocks, dropping the leap seconds that are now sometimes added that match the time shown on your watch with the Earth's rotation. Despite more than a decade of consideration, scientists have not reached a consensus about whether or not to make this change.

To go from UTC to the following time zones, subtract the numerical value of the hours shown (standard time/daylight saving time from the second Sunday in March to the first Sunday in November): Atlantic (−4/−3); eastern (−5/−4); central (−6/−5); mountain (−7/−6); Pacific (−8/−7); Alaska (−9/−8); Hawaii (−10).

CALENDARS

Every time the Earth goes around the sun, a year passes. This process takes about 365.25 days. Thus, after a year, the Earth has turned an extra ¼ turn than a mere 365 rotations. After four years, the Earth has completed one whole extra rotation—an extra day. We thus add an extra day to our calendars every fourth year, making it a leap year. Ordinarily, years that are divisible by four—2016, 2020, 2024—are leap years.

The system was set up by Julius Caesar, who named July after himself. Augustus Caesar named August after himself, and took a day from February to make August as long as July. At that time, the year began in March, which explains why September, October, November, and December come from the roots of the Latin words for 7, 8, 9, and 10.

A year is actually slightly shorter than 365.25 days: it is 365.2422 days. To keep from getting ahead of ourselves, we now omit the leap year every hundredth year. Thus 2100, 2200, and 2300 will not be leap years.

A year is defined from the interval between passages of the sun by the *vernal (spring) equinox*, one of the two points at which the ecliptic (see p. 17) and the celestial equator (see p. 6) cross. The *autumnal (fall) equinox* is the other crossing. The word *equinox* means equal night, but the lengths of day and night are not actually equal at the spring and fall equinoxes. For one thing, the Earth's atmosphere bends light from the sun, making it appear to rise when it is really still a few degrees below the horizon. Also, the equal lengths of day and night would apply to the time when the center of the sun rises, but daylight starts when the top of the sun rises. Thus, the day is about 10 minutes longer than the night at the equinoxes, and equal day and night precede the vernal equinox and follow the autumnal equinox by a few days.

Acknowledgments

I thank Naomi Pasachoff, Liz Stell, Marian Warren, Anne T. Pasachoff, Eloise Pasachoff, Deborah Pasachoff, and Eric Kutner for reading drafts of the first edition, and Madeline Kennedy and Michele Rech for assistance with the second edition. I thank Harry Foster and Barbara Stratton for their editing on the first edition, and Lisa White, director of guidebooks for Houghton Mifflin Harcourt, for continued work on updates for the first edition and for the second edition. I appreciate the work of Laney Everson and Beth Fuller, also at HMH, on the second edition.

Star maps by Wil Tirion (airbrush assistance on backgrounds by Wim van Dijk)
Constellation paintings by Robin Brickman

Image credits

5, 6, 10, 12, 58, 106 Jay M. Pasachoff Collection

7, 62 Image by J. C. Casado—Teide Observatory (IAC)—starry earth.com/TWAN (The World At Night)

11 Jimmy Westlake, Colorado Mountain College

15 Axel Mellinger (Central Michigan U.), A Color All-Sky Panorama Image of the Milky Way, *Publ. Astron. Soc. Pacific* **121**, 1180-1187 (2009).

17 NASA/JPL/DLR

46 ESO/Igor Chekalin

48 NASA/Chandra X-ray Center/Smithsonian Astrophysical Observatory

50, 70 Lorenzo Comolli, www.astrosurf.com/comolli

52 Nigel A. Sharp/NOAO/AURA/NSF

54 NASA and The Hubble Heritage Team (STScI/AURA)

56 NASA, ESA, and AURA/Caltech

60 Craig Stark

64 2MASS/NASA

66 ESO

71 NASA, ESA, and the Hubble Heritage (STScI/AURA)–ESA/Hubble Collaboration

72 NOAO/AURA/NSF

73 ESO/INAF-VST/OmegaCAM; A. Grado/INAF-Capodimonte Observatory

75 *Top:* B. J. Fulton for creating the image with data collected at the Byrne Observatory at Sedgwick, a Las Cumbres Observatory Global Telescope facility (optical data), and at the Palomar Observatory 48-inch Oschin telescope (H-alpha channel) *Bottom:* NASA, ESA, and P. Challis (Harvard-Smithsonian Center for Astrophysics)

76 NASA/ESA/J. Hester (Arizona State U.)

77 NASA/Chandra X-ray Center/M. Weiss

78 NASA, ESA, and the Hubble Heritage (STScI/AURA)–ESA/Hubble Collaboration

79, 98 NASA, ESA, S. Beckwith (STScI), and The Hubble Heritage Team (STScI/AURA)

80 NASA, ESA, and The Hubble Heritage Team (STScI/AURA); P. Knezek (WIYN)

81, cover ESO/B. Tafreshi (twanight.org)

83 NASA, ESA, G. Illingworth, D. Magee, and P. Oesch (University of California, Santa Cruz), R. Bouwens (Leiden University), and the HUDF09 Team

86 NASA/Johns Hopkins University Applied Physics Laboratory/Carnegie Institute of Science

87 Jay M. Pasachoff (Williams College) and Ronald Dantowitz (Clay Center Obs./Dexter Southfield School)/National Geographic Society Committee for Research and Exploration

88, 100, 102, 108, 115, 117, 119 Jay M. Pasachoff

89 NASA/JPL-Caltech/USGS

90–91 NASA/JPL-Caltech/Ken Kremer/Marco Di Lorenzo

92, 94, 95 NASA/JPL/Space Science Institute

93 NASA, ESA, H. Weaver and E. Smith (STScI) and J. Trauger and R. Evans (NASA's Jet Propulsion Laboratory-Caltech)

96 Imke de Pater (UC Berkeley) and Heidi Hammel (AURA)/W. M. Keck Observatory

97 NASA/ESA/Lawrence Sromovsky (U. Wisconsin, Madison), Heidi Hammel (AURA; then Space Science Institute), and Kathy Rages (SETI Inst.)

99 NASA/Johns Hopkins University Applied Physics Laboratory/Southwest Research Institute

101 ESO/Gabriel Brammer (now STScI)

102 B. Tafreshi (twanight.org)

105, 107 Ernest Wright, NASA's Goddard Space Flight Center/Scientific Visualization Studio

109 *Top:* Jason Morgan. *Middle:* Xavier Jubier. *Bottom:* Michael Zeiler

110 NASA's Ames Research Center

111 NASA

113 *Top:* P. Goode, W. Cao, and Big Bear Solar Observatory/New Jersey Institute of Technology. *Middle:* Royal Observatory Belgium/ESA, courtesy of Daniel B. Seaton; (c) NASA/LMSAL/Solar Dynamics Observatory. *Bottom:* NASA/LMSAL/Solar Dynamics Observatory

116 *Top:* images by Ronald Dantowitz and Nicholas Weber (Clay Center Obs./Dexter-Southfield Schools) and Jay M. Pasachoff, composited by Pavlos Gaintatzis, Aristotle University of Thessaloniki, Greece. *Bottom:* images by Jay M. Pasachoff; composited by Muzhou Lu

120 Deborah Pasachoff

INDEX

Page numbers in *italic* refer to illustrations.